Numbers, Sequences and Series

Modular Mathematics Series

Numbers, Sequences and Series

Keith E Hirst

Faculty of Mathematical Studies,
University of Southampton

Edward Arnold
A member of the Hodder Headline Group
LONDON   MELBOURNE   AUCKLAND

First published in Great Britain 1995 by
Edward Arnold, a division of Hodder Headline PLC,
338 Euston Road, London NW1 3BH

British Library Cataloguing in Publication Data
A catalogue record for this book is available from the British Library

ISBN 0 340 61043 3

1 2 3 4 5   95 96 97 98 99

Typeset in 10/13 Times by
Paston Press Ltd, Norfolk
Printed and bound in Great Britain by
J W Arrowsmith Ltd, Bristol

Contents

Series Preface

This series is designed particularly, but not exclusively, for students reading degree programmes based on semester-long modules. Each text will cover the essential core of an area of mathematics and lay the foundation for further study in that area. Some texts may include more material than can be comfortably covered in a single module, the intention there being that the topics to be studied can be selected to meet the needs of the student. Historical contexts, real life situations, and linkages with other areas of mathematics and more advanced topics are included. Traditional worked examples and exercises are augmented by more open-ended exercises and tutorial problems suitable for group work or self-study. Where appropriate, the use of computer packages is encouraged. The first level texts assume only the A-level core curriculum.

Professor Chris D. Collinson
Dr Johnston Anderson
Mr Peter Holmes

Preface

Numbers and geometry are historically the two foundations upon which mathematics has been built over some 3000 years, and the subjects discussed in this book span much of that period. We consider some topics which were studied by the Pythagorean school of philosophy in Ancient Greek times around 500–600 BC, and also ideas about numbers which were formulated at the end of the 19th century.

After an introductory chapter, which develops some of the logical structures used in mathematical reasoning, there are five chapters dealing with different features of the number system. I have not attempted to give a mathematically complete development of number systems right from their basic logical foundations, but to document some of the important and more interesting and applicable aspects. I have chosen the material so as to illustrate mathematical techniques which are used in other contexts. So, in Chapter 2, which explores integers, we begin to consider the method of making deductions from a basic set of rules or axioms, a method which has become the hallmark of many parts of mathematics during the past couple of centuries. We also consider algorithms, which are systematized procedures for carrying out calculations. These occur in the work of Euclid around 300 BC but are of particular importance now that they can be implemented on computers. In Chapter 3 we apply the notion of an equivalence relation introduced in Chapter 1 to the study of fractions. Again this tool is used throughout mathematics. One of the Pythagorean discoveries was related to the notion that there is no fraction whose square is equal to 2. Describing mathematically how to complete the system of fractions so as to include numbers like square roots was one of the achievements of the late 19th century, and in Chapter 5 I have chosen one of several approaches, again because of the widely applicable ideas involved. Finally, we look briefly at the complex numbers, where negative numbers have square roots.

Having laid the foundations of the number system we turn to the analysis of infinite processes involving sequences and series of numbers. The methods and concepts are those which help to provide the mathematical foundations of the theory of limiting procedures, which in turn lie at the heart of the differential and integral calculus. Chapter 9 uses this material to give an account of the system of decimal representation of numbers.

The final chapter of the book gives a glimpse of some of the directions in which the topics studied may be developed, and includes some suggestions for self-study projects.

Throughout the book I have emphasized the multiple perspectives necessary to gain a good understanding of any area of mathematics. These are exemplified in Chapter 4 on inequalities, where I have exhibited the notion of proof by logical deduction, and the illustration of concepts and techniques from a graphical, numerical and algebraic viewpoint. The notion of proof based on logical inference is central to mathematics and I have given some prominence to 'indirect' methods of proof involving logical contradiction, and also to proof by Mathematical Induction, as codified by Guiseppe Peano in 1889.

Most authors of textbooks share the view that mathematics cannot be learned simply by reading, but that there has to be some active participation, consequently the worked examples and the exercises are a vital component of this book. In addition, sharing ideas with others is indispensable, and I have included a number of tutorial problems aimed at stimulating group work and discussion, either in class time or through private study.

Computers impinge on most areas of mathematics nowadays, both for their numerical and their graphical capabilities. I have included small segments of computer code where it is appropriate, especially to illustrate the workings of some algorithms. The book can be read without this component however. Equally important is the geometrical representation of sequences through their interaction with graphs. There are many excellent packages available, and of these the one I have chosen to refer to explicitly is *A Graphic Approach to Calculus* (abbreviated in the book as Graphical Calculus) by David Tall and Piet van Blokland, available through Rivendell Software, 21 Laburnum Avenue, Kenilworth, Warwickshire, CV8 2DR, UK. This runs on an IBM compatible PC with CGA, EGA or Hercules graphics. For readers with access to a BBC or BBC Master computer there are earlier versions under the title of *Supergraph*. There are, of course, some public domain graph plotters available.

Finally, there are a number of acknowledgements. I am grateful to the series editor, Dr Johnston Anderson, and to the publishers for their efforts in helping this book to come to fruition. I am also grateful to my colleagues at Southampton—in particular Professor Martin Dunwoody—for the interest they have shown. I am grateful to Ray d'Inverno and David Firth for introducing me to the intricacies of T_EX and P_ICT_EX which were used to produce the manuscript and diagrams.

Without our students the drive to communicate mathematics would be considerably less, and in this context I am most grateful to Gemma Cotterell, who read much of the book in draft and commented perceptively on many aspects. Finally , I have to acknowledge the help and support of my wife, Ann, without whom nothing would be possible.

Keith Hirst, Southampton
December 1993

1 • Sets and Logic

Sets are the building blocks of mathematics. We use them to construct abstract ideas from simpler ideas or objects. For example, the idea of 'even' can be thought of as a common property abstracted from the set of numbers $\{2, 4, 6, 8, \ldots\}$. The idea of a continuous function is an abstraction of a common observation about the set of graphs of many familiar functions. In this case it is much more difficult to imagine a list of all the continuous functions, and an alternative way of describing a set is through some defining property which all the members of the set share. So our set of even numbers can be described as the set of all positive whole numbers that are divisible by 2 with no remainder. In the case of continuous functions it is more difficult to give such a property, and indeed the formulation of a definition of a continuous function occupied mathematicians for a good part of the 18th and 19th centuries.

If sets are the building blocks, then logic provides a framework whereby these building blocks can be joined together to form mathematical theories. Logic provides agreed criteria of validity whereby we can establish properties within a theory in the way that Euclid did in his books known as *The Elements*, compiled in about 300 BC. The study of logic and the formulation of some of the logical rules of deduction were carried on in ancient Greece, particularly by Plato and Aristotle during the 4th century BC, and this provided models for reasoning, not only in mathematics.

In the late 19th and early 20th centuries, there was a resurgence of interest in the study of logic as a subject. One of the aims of the time was to try to establish a foundation for the whole of mathematics based purely on sets and logic. Attempts were made to give precise rules, in symbolic form, called axioms, for the theory of sets, and to do the same for symbolic logic. The enterprise was not as successful as its originators had hoped, but it gave rise to a wealth of ideas, some of which are particularly useful in some branches of Computer Science today.

In this chapter I shall not try to give an axiomatic description of sets and logic. I shall describe those aspects of these topics which will be helpful in the study of numbers and sequences, and introduce just the symbolism which is generally found to be useful in analysing some of the ideas involved.

1.1 Symbolism

Notation and symbolism in mathematics is often thought of as a barrier to understanding. This is not surprising as it emphasizes the abstract and general nature of some of the ideas and operations. It is important therefore to emphasize the functions of symbolism. Firstly, symbols can act as abbreviations for phrases that would be cumbersome if used in words repeatedly. Many of the common symbols are like this, and will be familiar from school mathematics. For example nobody these days would write equations with the word 'equals' in place of the symbol '=' which represents the notion of equality. However, it should be recalled that even such a common symbol had to have its first appearance in the literature. In this case it is usually ascribed to Robert Recorde in his book *Whetstone of Witte* (1557). He explains that he will use 'a paire of paralleles ... because noe 2 thynges can be moare equalle'. The parallel lines in his symbol for equality are much longer in fact than we use today. Abbreviation is only one function however. The main use of symbols is when they relate to one another, as in an equation or in a formal statement of some kind. The aim there is to try to eliminate the ambiguity which is bound to occur when verbal statements are used, just by the nature of human language. For example, the statements 'You do that again and you will be in trouble' and 'If you do that again then you will be in trouble', which we would use synonymously, suggest that in ordinary language the word 'and' can sometimes be used in place of 'if ... then'. This is certainly not the case in logical deductions, and the use of some form of symbolism can help to emphasize that we are using the language of mathematics rather than the language of English discourse. The structure of a symbolic statement can often help to indicate similarities between disparate ideas, in a way that verbal statements would not. It can show us for example that the logical structure of two mathematical arguments is identical, even though the content of one may be arithmetic, whereas the other could concern geometry. We shall use symbolic statements when they are helpful, while aiming to keep them at a level which does not inhibit understanding of the underlying ideas. In many cases we shall use both verbal and symbolic forms, and discuss the translation between them. In fact, it is often helpful to understanding to translate a symbolic statement into words.

We will be discussing various sets of numbers a great deal, and so we shall use a conventional symbolic letter to replace their verbal name, as follows: $\mathbb{N}$ will stand for the Natural Numbers (the counting numbers 1,2,3, etc); $\mathbb{Z}$ will stand for the Integers (whole numbers, positive, negative or zero—Z stands for the German word *Zahlen*, meaning simply 'numbers'); $\mathbb{Q}$ stands for the rational numbers, the fractions (Q stands for quotient, and a fraction is the quotient of two integers); $\mathbb{R}$ stands for the real numbers, all the numbers both rational and non-rational, including familiar non-rational numbers like $\sqrt{2}$ and π; finally, $\mathbb{C}$ will stand for the complex numbers (which include numbers whose squares are negative).

1.2 Sets

We have seen examples in the introduction to this chapter of sets described by lists and sets described by rules or properties. The aspect of a set of objects we are concerned with at this point is simply that of determining whether a given object is a member of that set. Imagine that we ask a class of 100 students to think of a number. We record each number on a piece of paper. We now simply want to test whether a given number, say 327, was thought of. We look in the list, and it is immaterial where 327 occurs. The order within the list therefore does not matter, and neither does it matter whether 327 appears more than once. Sometimes order does matter, as in coordinate geometry, where (1,2) and (2,1) are different points. In the basic theory of sets however, we are not concerned with order or repetition. These notions can be added later as developments of the theory when we need them.

We use the following notation in connection with the two ways of describing sets. We conventionally enclose sets within braces { }, and for sets described by lists we just list the members inside the braces, either as a complete list such as $S = \{2, 3, 5, 7, 11, 13, 17, 19\}$ where the role of S is simply to act as a name for the set in case we want to refer to it a great deal. Sometimes we cannot list all the members explicitly, and then we make assumptions about the reader interpreting the pattern we are trying to indicate in the way we intend. For example, if we write $T = \{3, 6, 9, 12, 15, \ldots\}$ we might assume that readers will interpret this as continuing and including all the positive integer multiples of 3. Using the same two sets we can introduce the notation for sets described by rules. So we have $S = \{x : x$ is a prime number less than 20$\}$ and $T = \{n : n = 3k$ for some positive integer $k\}$. Membership is indicated by the symbol $\in$, read as 'belongs to' or 'is a member of'. So we can write $7 \in S$ and $36 \in T$. For the negation, we would write $4 \notin S$ and $23 \notin T$ for example. We sometimes want to indicate what overall set of numbers some variable is to be drawn from in describing a set, so that, for example, we would write $\{t \in \mathbb{Q} : 2 \leq t \leq 3\}$, and read it as 'the set of rational numbers lying between 2 and 3 inclusive'. We will also need to translate from one form of description to another, as with the two descriptions of S and T above. As another example, we could write the fact that the numbers 2 and 3 are the only solutions of the quadratic equation $x^2 - 5x + 6 = 0$ as an equality between sets described in the two forms, as

$$\{x : x^2 - 5x + 6 = 0\} = \{2, 3\}.$$

EXERCISES 1.2

1. Write the following sets as lists:

   (i) $\{x \in \mathbb{N} : 2 < x^2 < 75\}$,

   (ii) $\{t : t \in \mathbb{Z}$ and $-4 \leq t \leq 4\}$,

   (iii) $\{p : p$ is a two digit prime number$\}$,

   (iv) $\{x : x^3 - 2x^2 - 5x + 6 = 0\}$,

   (v) $\{y : y^3 - 2y^2 + y = 0\}$.

2. Write the following sets using rules:

   (i) $\{1, -1, 2, -2, 3, -3\}$,

   (ii) $\{1, 4, 7, 10, 13, 16, 19, 22, \ldots\}$,

   (iii) $\{-1, -4, -9, -16, -25, -36, \ldots\}$,

  (iv) $\{0.1, 0.01, 0.001, 0.0001, 0.00001, \ldots\}$,

  (v) $\{a, e, i, o, u\}$.

3. Decide which of the following are true statements:

   (i) $6 \in \{2, 4, 8, 16, 32\}$,

   (ii) $10 \in \{x : x = 3n + 1 \text{ and } n \in \mathbb{N}\}$,

   (iii) $\{2\} = \{x : x^2 = 4\}$, (notice the notation $\{2\}$ for the set containing only one member, the number 2),

   (iv) $\{1, 2, 1, 4, 1, 6, 6\} = \{1, 6, 4, 2\}$,

   (v) $\{p, q, r, s\} = \{p, q, \{r, s\}\}$,

   (vi) $\{2w : w \in \mathbb{Z}\} = \{2, 4, 6, 8, 10, \ldots\}$.

1.3 The Logic of Mathematical Discourse

The truth of a mathematical statement is investigated by two means. The first involves analysing the statement into its constituent parts and understanding the logical connections between them. The second involves analysing the mathematical content of the components themselves. The tools used in the second case will be those of algebra, geometry, calculus etc. In the first type of analysis, the main tool is that of logic, and we shall discuss some aspects of that in this section. It is not my intention to explore mathematical logic in detail, but to understand the ways in which simple connecting words and phrases are used to relate statements. The simplest of these is 'and'. For a statement like '3 is prime and 4 is a perfect square', its truth relies on the fact that we understand the use of the word 'and' to require that both components of the statement are true, and indeed our knowledge of numbers tells us that this is the case. The statement '3 is prime and 6 is an odd number' will be seen to be false by the same usage of the word 'and'. With this particular connective word there is no ambiguity. The word 'or' does give rise to ambiguity, because in English it is used sometimes in an inclusive sense and sometimes in an exclusive sense. If, in English, it is important to be precise we sometimes use and/or for the inclusive case, and either–or for the exclusive case. In mathematical discourse it is conventional to use 'or' with the inclusive meaning, so that '3 is prime or 6 is an odd number' is taken as true, and also '3 is prime or 4 is a perfect square' is true, where the first compound statement has only one true component whereas the second compound statement has two true components. If we wish to use the exclusive 'or' then we would use a construction like 'P or Q but not both', so that '3 is prime or 6 is odd but not both' is true, whereas '3 is prime or 4 is a perfect square but not both' is false. Notice that the word 'but' has been used here. It carries a nuance of meaning in English different from 'and', whereas in mathematical discourse the logical function of 'but' is identical with that of 'and'.

Many mathematical arguments proceed by a sequence of deductions, which tell us that one statement in a chain of reasoning implies the next. The logical analysis of statements of implication is therefore important, and we shall discuss this is some detail. There are sets of synonymous words and phrases used in this context and we shall try to consider them all. Mathematical logic can be approached in a purely symbolic fashion, which we are not doing here, but it is convenient to use the abbreviation P⇒Q for the statement 'P implies Q'. The first task is to decide about the truth of the compound statement 'P implies Q' in relation to the truth of its constituent parts P (called the premise) and Q (the conclusion). If both are true then 'P implies Q' will be true. We should notice that the implication being considered here is a purely logical one; there need be no material connection between P and Q, so that the statement '2 is a prime number implies that this book has more that 10 pages' will be taken as logically true. If P is true and Q is false we cannot have P implying Q and so the compound statement 'P implies Q' will be false.

So far this reflects the use of implication in ordinary language, but now we have to grapple with the situation where P itself is false. The first reaction is to wonder how a false statement could imply anything. Recall however that in algebra we often work with equations which are true for some values of the variable and false for others, but we operate on them using valid rules of algebra. So, for example, we would agree on the truth of the statement '$x = y \Rightarrow x^2 = y^2$' in ordinary algebra, even though a special case would be '$-1 = 1 \Rightarrow 1 = 1$' where the premise is false but the conclusion is true. A non-mathematical example will help to make the point. Consider the statement 'if you pass the examination I will give you £50'. The only circumstance in which you might accuse me of lying is if you *did* pass the examination and I *did not* give you £50. If you didn't pass and did not get £50 you would not be aggrieved. If you did not pass but I nevertheless gave you £50 you might be surprised, but you would not claim to have been cheated. We often write 'if P then Q' for 'P implies Q' and the if–then phraseology may help to make it clearer that we are dealing with conditional statements. Acceptance of the algebraic statement concerning squaring also entails the truth of '$2 = 5$ implies $4 = 25$', which we accept as logically true even though both components are false. We emphasize that the reason is not that we habitually use such arithmetic statements, but that we often use conditional statements involving variables, whose truth is contingent on the values of these variables. There follows a table of synonyms for implication.

P⇒Q	Q⇒P
P implies Q	Q implies P
if P then Q	if Q then P
Q if P	P if Q
P only if Q	Q only if P
P is sufficient for Q	Q is sufficient for P
Q is necessary for P	P is necessary for Q
P entails Q	Q entails P
Q follows from P	P follows from Q

As one example, the equivalence of 'P implies Q' and 'P is a sufficient condition for Q' reminds us that a careful analysis of the logic of both statements tells us that each is saying that knowledge of P enables us to deduce Q. Finally, from the discussion above, we emphasize that if we are trying to prove a statement of the form 'P implies Q' the only way it can fail is if P is true and Q is false.

Care is needed to distinguish between 'if' and 'only if'. Consider the statement 'the number n is divisible by 4 *if* its units digit is even'. This can be re-phrased 'if the units digit of the number n is even then n is divisible by 4'. This is false, as the example $n = 42$ shows. Now consider the statement 'the number n is divisible by 4 *only if* its units digit is even'. This is equivalent to saying 'if the number n is divisible by 4 then its units digit must be even'. This is true, since if n is divisible by 4 its units digit could certainly not be odd.

TUTORIAL PROBLEM I

Discuss the synonyms in the table above, finding examples both from everyday language and from mathematical discourse which help to clarify the logical equivalences involved.

Another phrase that is commonly used in mathematics is 'P if and only if Q'. This can be analysed as 'P implies Q and Q implies P', and in many situations where we want to prove an 'if and only if' statement we will prove the two implications separately. If we have a situation where 'P if and only if Q' is true we say that P and Q are logically equivalent.

The remaining logical idea we have to discuss in this section is that of negation, and particularly its interaction with implication. For single statements it is clear that 'not P' is true if and only if P itself is false. By analysing the logic of ordinary language it can be seen that 'not (P and Q)' is equivalent to '(not P) or (not Q)', with a reminder that 'or' is being used inclusively.

TUTORIAL PROBLEM 2

Find examples to illustrate the equivalence of 'not (P and Q)' with '(not P) or (not Q)'. Illustrate also the equivalence of 'not (P or Q)' with '(not P) and (not Q)'. Investigate the analogous equivalence if 'or' were used in the exclusive sense.

To analyse the interaction of negation and implication, we will start with an example.

Example I

Show that if m^2 is an even number then m is an even number.

Let us suppose that m is an odd number. Then m can be written in the form $m = 2t + 1$, where t is an integer. Squaring then gives

$$m^2 = (2t + 1)^2 = 4t^2 + 4t + 1 = 2(2t^2 + 2t) + 1,$$

showing that m^2 is odd. Common sense logic then argues as follows: if we have m^2 even then we could not have m odd, because that would imply that m^2 is odd. So m itself is in fact even.

The underlying logic of the argument in Example 1 can be expressed through the logical equivalence of the statement 'P implies Q' with '(not Q) implies (not P)'. This is exactly what we have used. We have shown that m^2 even implies m even by establishing the logically equivalent statement that m not even implies m^2 not even. This kind of argument occurs sufficiently often in proofs for us to introduce some nomenclature. If 'P implies Q' is a conditional statement, then its equivalent form 'not Q implies not P' is called the *contrapositive*.

Given a statement 'P implies Q' of implication, we often wish to consider the reverse implication 'Q implies P', called the *converse* of 'P implies Q'. It is important to realize that a statement and its converse are not logically equivalent. For instance, the ordinary rules of algebra give us the truth of '$a = b$ implies $a^2 = b^2$'. The converse is not true however, because for instance $(-2)^2 = (+2)^2$, but this does not imply that $-2 = +2$.

Sometimes the converse of a statement *is* true, but this is a consequence of the mathematical content of the statement and not a matter of logic. So, for example, the converse of the statement in Example 1 is true, because m even does imply that m^2 is even.

The argument in Example 1 is an illustration of an indirect proof, where one assumes that the conclusion is false. A related method of proof is known by the Latin term *reductio ad absurdum*. One of the best known examples of this is the proof that $\sqrt{2}$ is not a rational number, and this is given as Proposition 1 in Chapter 5. The logical idea is that we assume the conclusion to be false, and derive a logical contradiction saying that some other statement R is both true and false. This is absurd, and so the original assumption that the conclusion was false cannot be correct. Indirect proofs occur a great deal, and in solving problems it is always a good strategy to ask whether an indirect method might work if one cannot see a method involving a direct chain of deduction from premise to conclusion.

EXERCISES 1.3

1. Write the contrapositive and converse of the following statements of implication. Decide in each case whether or not the converse is true.

    (i) If $m^2 = 10$ then m is not a rational number.

    (ii) If $x = 2$ then $x^2 - 5x + 6 = 0$.

    (iii) If $\theta = \pi$ then $\sin \theta = 0$.

    (iv) If $f(x) > 0$ for all x between 0 and 1 then $\int_0^1 f(x) \, dx > 0$.

    (v) If $f'(a) = 0$ and $f''(a) < 0$ then $f(x)$ has a local maximum at $x = a$.

2. Give proofs of the following statements using the contrapositive. You may wish to rephrase them, referring to the logical equivalences discussed before Tutorial Problem 1 above.

   (i) The number m being odd is a necessary condition for m^3 to be odd.

   (ii) $t = 1 + 2 + \ldots + k$ if $t = k(k+1)/2$.

   (iii) $x^2 < 2$ is a sufficient condition for $x^2 + x - 12 < 0$.

   (iv) The integer n cannot be expressed as a sum of two square integers if n has remainder 3 on division by 4.

   (v) A chord of a circle does not pass through the centre of the circle only if it subtends an angle different from 90° at the circumference.

1.4 Statements Involving Variables

The statement '$x^2 - 2x - 3 = 0$' is neither true nor false. It contains the 'free variable' x. When we are discussing such a statement we have to make it clear what set values of x can be drawn from. We are often interested in finding the values of x for which the statement is true, and in this case, if x can stand for any integer, it is true when $x = -1$ and when $x = 3$. On many occasions however, an equation or some other kind of statement involving a variable may be too complicated to solve, and we will then be interested to know whether there are any values of x at all which when substituted will give a true statement. Statements involving variables are sometimes referred to as open statements or open sentences, to distinguish them from statements such as '13 is a prime number' which are definitely either true or false.

Sometimes we find statements which are true for all values of x which can be substituted: for example, $x^2 - 1 = (x - 1)(x + 1)$ is true for all numbers x. Another example is $\cos^2\theta + \sin^2\theta = 1$, again true for all numbers θ. Such statements are often called identities.

So the two most common questions we ask about a statement involving a variable are (i) is there any value of the variable which makes the statement true? (ii) is the statement true for all values of the variable? It is useful to symbolize such statements, as they appear in many places later in this book and throughout mathematics.

The symbolic statement $\exists x \in T, S(x)$ stands for 'there exists a member x of the set T such that $S(x)$', where $S(x)$ is some statement involving the variable x. So, for example, as we saw at the beginning of this section, the statement $\exists x \in \mathbb{Z}, x^2 - 2x - 3 = 0$ is a true statement. To demonstrate its truth we can specify those values of x which satisfy the equation, namely $x = -1$ and $x = 3$. Notice however, that to demonstrate the existence we only need to produce one value of x which satisfies the equation. Finding both is superfluous, but of course more informative.

The symbolic statement $\forall x \in T, S(x)$ stands for 'for all (for each, for every) member(s) x of the set T, $S(x)$', where $S(x)$ is some statement involving the variable x. The two identities given earlier in this section illustrate this, for the statements

$$\forall x \in \mathbb{R}, x^2 - 1 = (x-1)(x+1) \quad \text{and} \quad \forall \theta \in \mathbb{R}, \cos^2\theta + \sin^2\theta = 1$$

are both true. To demonstrate their truth is part of the development of elementary algebra and trigonometry respectively. Generally, to prove 'universal' statements of the form $\forall x, \ldots$ is more difficult than to prove 'existential' statements $\exists x, \ldots$. Statements of the kind we have been considering are often referred to as *quantified* statements. The symbol $\forall$ is the *universal quantifier*, and $\exists$ is the *existential quantifier*. It is important to note that in a quantified statement the variable is no longer a *free* variable. This means that, for example, the two statements

$$\forall x \in \mathbb{R}, x^2 - 1 = (x-1)(x+1) \quad \text{and} \quad \forall t \in \mathbb{R}, t^2 - 1 = (t-1)(t+1)$$

say exactly the same thing. The letter used for the variable is immaterial.

Throughout this book we shall be working with statements involving quantifiers. Their expression in symbolic form does offer some clarity, but it is important to be able to handle both symbolic and verbal forms, and we shall work with both in most cases. So, for example, if we consider a statement like 'all positive real numbers have a real square root', we can see that it begins 'all positive real numbers'. We have a choice as to how to express this, for example we could define $\mathbb{R}^+$ as a symbol for the positive reals and write '$\forall x \in \mathbb{R}^+$', or we could take it as understood that x is real, and write '$\forall x > 0$', or we could write '$\forall x \in \mathbb{R}, x > 0 \Rightarrow \ldots$'. Now the word 'have' in the sentence is asserting the existence of a number with the specified property. We shall need to choose a variable to stand for this number, say y. To say that y is a square root of x is equivalent to saying that $y^2 = x$. So we can reformulate the verbal statement in an alternative verbal form which corresponds more closely to the analysis we have undertaken; 'for every positive real number x, there exists a real number y such that $y^2 = x$'. Finally, we can translate this into its symbolic counterpart;

$$\forall x \in \mathbb{R}^+, \exists y \in \mathbb{R}, y^2 = x.$$

When we encounter such statements we shall sometimes want to disprove them, and this requires us to think about the negation of a quantified statement. We shall do this through an analysis of ordinary language. If we say that a statement of the form 'for all x, $S(x)$' is false, this is logically equivalent to saying that it is possible to find at least one example of a value of x which renders $S(x)$ false. So, for example, the statement 'for all integers x, $x/2$ is an integer' is shown to be false merely by saying that if $x = 3$ then $x/2$ is not an integer. Such a case is known as a *counterexample*. If we wish to negate a statement of the form 'there exists x such that $S(x)$' then we have to demonstrate that for all values of x, $S(x)$ itself is false. To summarize:

$\text{not}(\forall x, S(x))$ is equivalent to $\exists x, (\text{not } S(x))$,

$\text{not}(\exists x, S(x))$ is equivalent to $\forall x, (\text{not } S(x))$.

EXERCISES 1.4

1. Translate the following verbal statements into symbolic statements using quantifiers. In each case say whether the statement is true.

   (i) Every integer is a rational number.

   (ii) There is a rational number between -1 and $+1$.

   (iii) All the roots of the equation $2x^2 - 5x + 3 = 0$ are integers.

   (iv) Every rational number is less than its square.

   (v) There is a real number satisfying $\sin x = -2$.

2. Express each of the following symbolic statements in verbal form, and state whether each is true. Write the negation of those statements that are false using quantifiers.

   (i) $\forall x \in \mathbb{R}, \cos x = 0 \Rightarrow |x| < 2\pi$.

   (ii) $\exists q \in \mathbb{Q}, 23q = 78$.

   (iii) $\forall h \in \{1, 2, 3\}, h^3 - h + 7 > 0$.

   (iv) $\forall x \in \mathbb{R}, \cos \pi x = 0 \Rightarrow x \in \mathbb{Q}$.

   (v) $\forall t \in \mathbb{Z}, t < 0$ or $\sqrt{t} \in \mathbb{R}$.

1.5 Statements Involving More Than One Variable

The statement $\forall x \in \mathbb{R}^+, \exists y \in \mathbb{R}, y^2 = x$ about square roots which we encountered near the end of the previous section involved two quantifiers, and we need to consider such statements a little more closely. Consider the two statements 'every citizen has a duty to help others' and 'every citizen has a date of birth'. In the first case the abstract notion of 'duty to help others' is something which does not vary from one citizen to another (although the manner in which that duty is discharged may vary). On the other hand the 'date of birth' is something that will vary from one person to another. We shall find it important to distinguish these cases, and the use of quantifiers will help to clarify this distinction. Let us consider the statement about birthdays. We can reformulate this using variables as follows: for every person P, there is a date D such that D is P's birthday. We would all agree that such a statement is true. If we now reverse the order of quantification we obtain the statement: there is a date D such that, for every person P, D is P's birthday. Is this true or false? To decide this we have to try to produce a value of D which makes the statement [for every person P, D is P's birthday] true. Let us try one value, for example D = 4th February. The statement in brackets then becomes [for every person P, 4th February is P's birthday]. This is patently false, and I am sure everyone could provide a counterexample. Any other value of D we try will similarly render the statement false. This demonstrates that the order of quantification is important, whereas sometimes in ordinary language the order can be used ambiguously. Let us now consider a more mathematical example with the two statements:

$$\forall x \in \mathbb{Z}, [\exists y \in \mathbb{Z}, x < y] \quad \text{and} \quad \exists y \in \mathbb{Z}, [\forall x \in \mathbb{Z}, x < y].$$

We do not usually find square brackets used in such statements; they are inserted here for clarity of explanation. The first statement is asserting that, given any real number x, there is a larger one y. This is clearly true, for example $y = x + 2$ will suffice. It should be noted here that the value of y can and does depend upon x. The second statement is asserting the existence of a real number greater than all real numbers, and this is false; there is no largest real number. A further analysis using negation of quantified statements will serve to emphasize this conclusion. The statement

$$\text{not}(\exists y \in \mathbb{Z}, [\forall x \in \mathbb{Z}, x < y])$$

is equivalent to

$$\forall y \in \mathbb{Z}, \text{not}[\forall x \in \mathbb{Z}, x < y],$$

which in turn is equivalent to

$$\forall y \in \mathbb{Z}, \exists x \in \mathbb{Z}, \text{not}[x < y],$$

and therefore equivalent to

$$\forall y \in \mathbb{Z}, \exists x \in \mathbb{Z}, x \geq y,$$

which is easily seen to be true, for example by taking $x = y + 1$ (or even $x = y$).

TUTORIAL PROBLEM 3

Find some more statements, both from ordinary language and from elementary mathematics, where changing the order of quantification changes the meaning and truth of such statements.

TUTORIAL PROBLEM 4

Discuss the equivalence between the symbolic statement

$$\forall a > 0, \exists b > 0, b < a$$

and the verbal statement 'there is no smallest positive real number'. Illustrate why this is true by using a number line.

Formulate an analogous symbolic statement equivalent to 'there is no largest real number'.

EXERCISES 1.5

1. Translate the following verbal statements into symbolic statements using quantifiers. In each case say whether the statement is true.

    (i) There is an odd integer which is an integer power of 3.

    (ii) Given any positive rational number, there is always a smaller positive rational number.

    (iii) Given a real number a, we can always find a solution of the equation $\cos(ax) = 0$.

    (iv) For every real number x we can find an integer n between x and $2x$.

    (v) Given any real number k there is a solution of the equation $kx = 1$.

    (vi) For every positive real number a there are two different solutions of the equation $x^2 = a$.

2. Express each of the following symbolic statements in verbal form, and state whether each is true. Write the negation of those statements that are false using quantifiers.

    (i) $\forall x \in \mathbb{R}, \exists y \in \mathbb{R}, x + y = 0$.

    (ii) $\exists y \in \mathbb{R}, \forall x \in \mathbb{R}, x + y = 0$.

    (iii) $\forall t \in \mathbb{R}, \forall n \in \mathbb{N}, nt > t$.

    (iv) $\exists a \in \mathbb{N}, \exists b \in \mathbb{Z}, a^2 - b^2 = 3$.

    (v) $\forall u \in \mathbb{R}, \exists v \in \mathbb{R}, \forall w \in \mathbb{R}, u + v \leq w$.

1.6 Equivalence Relations

Human beings classify their experiences. It is the only way to make sense of the world. Think about a young child learning colours. Parents will talk to the child about a variety of toys and other things and use the word 'red'. The child somehow has to discern what these phenomena have in common and to ignore differences in order to acquire the abstract concept of 'red'. The fact that one object is a lego brick and the other a pair of mittens must be ignored. It is the idea that they are both red that is being emphasized. The learning is experiential: you can't tell a 3 year old child about light of a certain wavelength!

We engage in this kind of activity all our lives; it is the basis of language. If we think about any everyday word like chair, car, umbrella, bird, we are generally thinking not of individual objects but of classes of objects. This is done also in science and mathematics. Biologists classify organisms, and indeed there is a whole subject of taxonomy. Organic chemists think about alcohols, a whole class of compounds which have some common feature. Physicists think about the concept of force, which manifests itself in a variety of circumstances. In school mathematics we learn the notion of a right-angled triangle, even though there are infinitely many different possible shapes and orientations. We learn to regard sets of equivalent fractions as relating to the same number.

In everyday life, classifications are often a bit fuzzy; many of you will have had arguments about whether a colour is blue, green or turquoise. Sometimes we have to be exact however; for example every motor vehicle must be precisely classified as a car, van, motorcycle, heavy goods vehicle, etc, for the purposes of road tax. Another example relates to hovercraft. When they were first developed it was very difficult to decide whether they should be covered by regulations for ships or for aircraft, and it was legally important to decide. It is this exact kind of classification we use in mathematics, and as with motor vehicles we divide up a set into subsets in a particular way. Let us pursue this example a bit further. Let V denote the set of all motor vehicles in the UK. This set is *partitioned* into subsets {car, van, ...} in the following way.

(i) Any two subsets are either identical or disjoint.

(ii) The subsets together make up the whole of V.

(iii) No subset is empty.

In mathematics, an alternative way of characterizing such partitions has been developed, by looking at relationships between individual objects in the same class, which is somewhat analogous to the way in which road tax regulations seek to define vehicle types by specifying characteristic properties that all vehicles of a particular class share. We consider relationships all the time in mathematics: for example, we might say that one number is *smaller than* another; one number is the *square of* another; one triangle is a *reflection of* another; one fraction is *equivalent to* another. Relationships arising from partitions are called *equivalence relations*. They are characterized by three properties described below. Before we give the abstract definitions let us look at another example. We consider the set $\mathbb{N}$ of natural numbers and say that a is equivalent to b if and only if a and b have the same remainder on division by 5. If we now group together numbers which are equivalent we will obtain five subsets:

$$\{1, 6, 11, 16, \ldots\}$$
$$\{2, 7, 12, 17, \ldots\}$$
$$\{3, 8, 13, 18, \ldots\}$$
$$\{4, 9, 14, 19, \ldots\}$$
$$\{5, 10, 15, 20, \ldots\}$$

You should check that these subsets satisfy the three properties of a partition explained earlier. The subsets of a partition are called *equivalence classes*, and so in this example there are five equivalence classes.

Following these examples we are now in a position to give definitions of partition and equivalence relation in terms of sets and logic.

● *Definition 1*

A partition of a set S is a collection $\{A, B, C, \ldots\}$ of subsets of S with the following properties.

(i) If K and H denote any two subsets of the partition then $K = H$ or K and H have no members in common (i.e. they are disjoint).

(ii) Every member of the subset S is a member of a set in the partition.

(iii) Every subset in the partition contains at least one member of S.

In defining an equivalence relation we shall need to introduce a symbol to stand for the relationship. We shall use the symbol $\equiv$. So, in the last example, $a \equiv b$ would stand for 'a has the same remainder on division by 5 as b'. In general, we would read $a \equiv b$ as 'a is equivalent to b' or 'a is related to b'. ●

● *Definition 2*

A relationship $\equiv$ on a set S is an equivalence relation if the following three properties are satisfied.

(i) (*reflexivity*) $\forall a \in S, a \equiv a$.

(ii) (*symmetry*) $\forall a, b \in S, a \equiv b \Rightarrow b \equiv a$.

(iii) (*transitivity*) $\forall a, b, c \in S, a \equiv b$ and $b \equiv c \Rightarrow a \equiv c$.

Verbally, these properties can be stated

(i) every a is related to itself,

(ii) if a is related to b then b is related to a,

(iii) if a is related to b and b is related to c then a is related to c. ●

We have quoted the example of equivalent fractions in this context, and we shall pursue that idea further in Chapter 3, using the notion of an equivalence relation. In the example involving remainders on division by 5, we saw how the relationship gave rise to a partition of the set of natural numbers.

TUTORIAL PROBLEM 5

Let S denote the set of lines in the plane. Show that the relationship described by 'l is parallel to m' is an equivalence relation on S. Discuss which of the three properties in Definition 2 are satisfied by the relationship described by 'l is perpendicular to m'.

We shall conclude this section by giving a proof that partitions and equivalence relations are really the same idea, looked at from different points of view.

● Proposition 1

Given an equivalence relation $\equiv$ on a set S, the collection of equivalence classes forms a partition of S.   ●

PROOF

We have to prove that the collection of equivalence classes satisfies the three properties in the definition of a partition.

Suppose A and C are any two equivalence classes which are not disjoint. So there is some member, say b, which belongs to both. Now let a denote any member of A and c any member of C. Since a and b both belong to A, we must have $a \equiv b$. Since b and c both belong to C we must have $b \equiv c$. Transitivity for equivalence relations now tells us that $a \equiv c$, and so a and c belong to the same equivalence class. Hence $A = C$.

Now let s denote any member of S. By reflexivity $s \equiv s$, and so every s is in some equivalence class, and likewise any equivalence class contains at least one member. This establishes the second and third properties of a partition.   ●

We have apparently not used the symmetry property of an equivalence relation in this proof, but in fact it is implicit in the first part. A purely symbolic argument would demonstrate this, but we have preferred a verbal proof, and symmetry is implicit in the language used.

● Proposition 2

Given a partition of a set S, the relation $\equiv$ defined by '$a \equiv b$ if and only if a and b belong to the same subset in the partition' is an equivalence relation.   ●

PROOF

Given any $a \in S$, then of course a is in the same subset as itself, and so $a \equiv a$. This proves reflexivity.

If $a \equiv b$ then a is in the same subset as b, so that naturally b is in the same subset as a, i.e. $b \equiv a$, proving symmetry.

Now suppose $a \equiv b$ and $b \equiv c$. So a is in the same subset as b and b is in the same subset as c. These two subsets are not disjoint, since b belongs to both, and so by the first partition property they must be identical. So a belongs to the same subset as c and hence $a \equiv c$, proving transitivity.   ●

The first two parts of this proof seem to be stating the obvious. This is because the notion of equivalence is so deeply embedded in our ordinary language.

● Example 2

A relationship is defined between integers by: $a \equiv b$ if and only if $a - b$ is an integer multiple of 4. Show that this is an equivalence relation and determine the equivalence classes.

Given any $a \in \mathbb{Z}$, $a - a = 0 \times 4$. This proves reflexivity.

If $a \equiv b$ then for some $n \in \mathbb{Z}$, $a - b = n \times 4$, so $b - a = (-n) \times 4$, showing that $b \equiv a$. This proves symmetry.

To demonstrate transitivity, suppose that $a \equiv b$ and $b \equiv c$, i.e. $a - b = n \times 4$ and $b - c = m \times 4$ for some $m, n \in \mathbb{Z}$. Adding the two equations gives $a - c = (n + m) \times 4$, i.e. $a \equiv c$.

To determine the equivalence classes, we note that $a - b = n \times 4$ can be rewritten as $a = b + n \times 4$, so that given a number b, adding any multiple of 4 will give a number in the same class as b. If we start with $b = 0$, then with $b = 1$ etc, we find that $b = 4$ gives the same class as $b = 0$, because 0 and 4 are equivalent. This gives just four equivalence classes, namely

$$\{\ldots, -8, -4, 0, 4, 8, 12, \ldots\}, \{\ldots, -7, -3, 1, 5, 9, \ldots\},$$

$$\{\ldots, -6, -2, 2, 6, 10, \ldots\}, \{\ldots, -5, -1, 3, 7, 11, \ldots\}. \qquad \bullet$$

EXERCISES 1.6

1.   For each of the following relations, determine which of the three properties of reflexivity, symmetry and transitivity they satisfy. For those which are equivalence relations describe the equivalence classes. (In each case the set is specified and then the definition of the relationship is given.)

(i) S is the set of citizens of Spain: $p \equiv q$ if and only if p is a brother of q.

(ii) S is the set of cities, towns and villages in England: $h \equiv k$ if and only if h is in the same county as k.

(iii) S is the set of triangles in the plane: $c \equiv d$ if and only if d can be obtained by a translation of c.

(iv) $S = \mathbb{Q}$: $x \equiv y$ if and only if $y = -x$.

(v) S is the set of differentiable real functions: $f(x) \equiv g(x)$ if and only if $f'(x) = g'(x)$ for all x.

(vi) S is the set of points in the plane: $(x_1, y_1) \equiv (x_2, y_2)$ if and only if $x_1^2 + y_1^2 = x_2^2 + y_2^2$.

Summary

There are two key ideas in this chapter, used throughout mathematics. The first is the process of mathematical proof, analysing the logical relationships between statements, especially those involving implications. The notion of indirect proof using the contrapositive or contradition is equally important. The other area emphasized is the logical analysis of quantified statements involving variables. This will be used in many places in this book, particularly in Chapters 5 and 7.

EXERCISES ON CHAPTER I

1. Write the following sets as lists:
   (i) $\{p : p$ is a prime number and $p^2 < 1000\}$,
   (ii) $\{(m, n) : m \in \mathbb{Z}, n \in \mathbb{Z}$ and $m^2 + n^2 \le 20\}$,
   (iii) $\{t \in \mathbb{R}: \tan t = -1\}$.

2. Write the following sets using rules
   (i) $\{1, 3, 6, 10, 15, 21, 28, \ldots\}$,
   (ii) $\{\ldots, -7\pi/2, -3\pi/2, \pi/2, 5\pi/2, 9\pi/2, \ldots\}$,
   (iii) $\{h,i,r,s,t\}$.

3. Decide which of the following are true statements:
   (i) $\{4, 3, 5, 9\} = \{9, 3, 4, 5\}$,
   (ii) $\{2\} = \{t : t$ is an even prime number$\}$,
   (iii) $\{n\pi : n \in \mathbb{Z}\} = \{x \in \mathbb{R} : \sin x = 0\}$,
   (iv) $-1 \in \{2x - 5 : x \in \mathbb{N}\}$.

4. Prove or disprove the following statements:
   (i) n is a multiple of 3 if the sum of its digits is a multiple of 3,
   (ii) n is a multiple of 3 only if the sum of its digits is a multiple of 3,
   (iii) n is divisible by 3 if the number formed from its last two digits is divisible by 3,
   (iv) n is divisible by 3 only if the number formed from its last two digits is divisible by 3,
   (v) n is divisible by 4 if and only if the number formed from its last two digits is divisible by 4.

5. Write the contrapositive and converse of the following statements of implication. Decide in each case whether the statement and/or the converse is true. Prove those statements and converses which are true, and give an explanation for those which are false.
   (i) If $m^2 = 9$ then $m = 3$.
   (ii) If $\theta = 3\pi/2$ then $\cos \theta = 0$.
   (iii) If $x = 2$ then $x^2 - 5x + 6 \ge 0$.

6. Translate the following verbal statements into symbolic statements using quantifiers. In each case say whether the statement is true.
   (i) The equation $\cos 2x = \cos^2 x - \sin^2 x$ is an identity.
   (ii) There are no rational numbers satisfying $x^2 - x - 1 = 0$.
   (iii) All the roots of the equation $x^3 - 4x^2 + 7x - 6 = 0$ are square integers.

7. Express each of the following symbolic statements in verbal form, and state whether each is true. Write the negation of those statements that are false using quantifiers.

   (i) $\forall c \in \mathbb{R}, c \neq 0 \Rightarrow \exists d \in \mathbb{R}, cd = 1$.

   (ii) $\forall t \in \mathbb{R}, \exists n \in \mathbb{Z}, n < t < n+1$.

   (iii) $\exists q \in \mathbb{Q}, \forall n \in \mathbb{Z}, n \times q \in \mathbb{N}$.

8. For each of the following relations, determine which of the three properties of reflexivity, symmetry and transitivity they satisfy. For those which are equivalence relations describe the equivalence classes. (In each case the set is specified and then the definition of the relationship is given.)

   (i) S is the set of people living in Southampton: $i \equiv j$ if and only if i is older than j.

   (ii) S is the set of points in the plane: $(x_1, y_1) \equiv (x_2, y_2)$ if and only if $x_1 + y_2 = x_2 + y_1$.

   (iii) $S = \mathbb{Z}$: $m \equiv n$ if and only if m is an integer multiple of n.

   (iv) $S = \mathbb{N}$: $a \equiv b$ if and only if $a - b$ is a prime number.

   (v) $S = \mathbb{Z}$: $c \equiv d$ if and only if $d = c + 7k$ for some $k \in \mathbb{Z}$.

$2 \bullet$ The Integers

'God made the integers, and all the rest is the work of man.' So wrote the mathematician Leopold Kronecker (1823–1891). Whilst we may find this statement strange, it does express the idea that any mathematical theory has to start somewhere; that there have to be some basic statements on which the theory is built. These statements are called axioms, and are not part of the development of the theory; they are agreed foundations. In the case of subjects like number theory and geometry, many of whose properties become familiar long before we wish to consider a formal theory, the axioms are formulated to express ideas that are part of our already existing knowledge and understanding. This activity was undertaken in respect of the natural numbers by Guiseppe Peano (1858–1932), and an analysis of his system, known as the Peano Axioms, reveals that they are based on the idea of counting; more precisely on the idea that each number has a successor. This notion is found useful in many situations where we have an ordered set of things, and the notion is included in some computer languages, for example Pascal.

Peano worked in a number of areas of mathematics, and was especially interested in mathematical logic. In 1889 he published a booklet entitled 'The principles of arithmetic, presented by a new method'. The bulk of this is written purely in mathematical and logical symbols, with the preface and a few explanatory notes in Latin. He refers particularly to the work of George Boole, Hermann Grassmann, Richard Dedekind and Georg Cantor as helping him to develop his axioms. Peano begins with nine axioms. Four of these are concerned with the use of the $=$ symbol, and essentially are saying that the relationship of equality has the properties of an equivalence relation, as defined in §1.6. The remaining axioms are used to give a formal characterization of the system $\mathbb{N}$ of natural numbers. Peano used the notation $a + 1$ to denote the number following a, and then used the axioms to define addition. This has an appearance of circularity, as though addition were being used to define addition. Later formulations dealt with this by using a separate notation for the number following a, and this helps to make it clear that the development is not circular. We shall follow this procedure.

2.1 Peano's Axioms

The system $\mathbb{N}$ of natural numbers satisfies the following axioms.

1. There is a successor function on $\mathbb{N}$, denoted by succ.
2. There is a member of $\mathbb{N}$, denoted by 1, which is not the successor of any number.
3. The successor function satisfies $\mathrm{succ}(a) = \mathrm{succ}(b)$ if and only if $a = b$, i.e. it is a one-to-one function.

4.   If S is a subset of $\mathbb{N}$ having the properties

   (i) $1 \in S$,

   (ii) if $k \in S$ then $\text{succ}(k) \in S$,

   then $S = \mathbb{N}$.

The last axiom is known as the axiom of induction. If one thinks of succession as reflecting the idea of counting on by 1, then this axiom says that if we start at 1 (property (i)) and continue counting on from wherever we reach (property (ii)), then we will eventually be able to reach any number. This sounds obvious, and the fact that it does demonstrates that we have the intuitive idea that we could count for ever, even though most of us will only ever have counted up to a few hundred perhaps. This is the aim of the axiomatization, to formulate a set of mathematical rules which faithfully reflect our views of what the number system is. Having achieved this formulation, it proves to give us an extremely powerful tool for establishing properties of the number system on a proper mathematical footing, often reflecting some pattern of regularity observed from a few special cases, and conjectured to be true in general.

We mentioned the use of the successor function in the Pascal programming language. The following segment of Pascal code would, in theory, print out all the natural numbers. (In practice of course most computer systems can only handle integers within certain limits.)

```
n:=1;
repeat
   write(n);
   n:=succ(n);      (this is equivalent to  n:=n+1)
until false;
```

Peano used his system of axioms to develop the whole of arithmetic, including the four arithmetic operations, rules of indices, divisibility and some elementary number theory, and some of the properties of rational and irrational numbers. We shall not give the full development here, but illustrate the methods used by explaining how addition can be defined from the axioms.

Addition is defined by the following two rules:

(i) $a + 1$ is defined to be the successor of a,

(ii) $a + (b + 1)$ is defined to be the successor of $a + b$.

The next important step is to prove that this does properly define addition of any two numbers. To do this we use the axiom of induction, as follows.

Let a denote an arbitrary member of $\mathbb{N}$. Let S be the subset of numbers b for which the two rules properly define $a + b$. Rule (i) for addition above tells us that $1 \in S$, while rule (ii) tells us that whenever $b \in S$ then the successor of $b \in S$. The axiom of induction then says that $S = \mathbb{N}$, so that $a + b$ is well defined for all numbers a and b.

The definition of addition is an example of definition by induction, sometimes called recursive definition. As another example of this, consider the following inductive definition of the factorial function.

We define $n!$ for $n \in \mathbb{N}$ by the following two properties,

(i) $1!$ is defined to be 1,

(ii) for all $k \in \mathbb{N}$, we define $(k+1)!$ to be $(k+1) \times k!$.

TUTORIAL PROBLEM I

Discuss how this gives a proper definition of the factorial function, using the axiom of induction.

As a further illustration we shall see how the Σ (Greek capital 'sigma') notation for sums can be fitted into the framework of inductive definition.

Suppose we have a function $f(i)$ defined on $\mathbb{N}$ (for example $f(i) = i^2$), and that we want to add a large number of successive values of this function. So, for example, we may want to find the sum of the first n square numbers. We use two forms of notation for this, expressed as the two sides of the identity

$$\sum_{i=1}^{n} i^2 = 1^2 + 2^2 + 3^2 + \ldots + n^2.$$

The right-hand side is meant to indicate the procedure of continual addition. The left-hand side is an abbreviated notation which makes use of the fact that we have a general formula (i^2) for the number we are adding, and is also often more algebraically convenient than the right-hand side notation. We can define the left-hand notation inductively as follows:

$$\sum_{i=1}^{n} f(i) \quad \text{is defined for all } n \text{ by} \quad \text{(i) } f(1) \text{ if } n = 1,$$

$$\text{(ii) } \sum_{i=1}^{n+1} f(i) = f(n+1) + \sum_{i=1}^{n} f(i).$$

Another way of appreciating the axiom of induction is to demonstrate its equivalence with an intuitively obvious property of the natural numbers, the well-ordering principle, which states that any non-empty subset of the natural numbers has a least member.

● *Proposition I*

The axiom of induction implies the well-ordering principle.   ●

PROOF

Let T be any non-empty subset of $\mathbb{N}$. If 1 is a member of T then it is the least member of T. So suppose that 1 is not a member of T. We now proceed to construct a proof by contradiction. Assume that T does not have a least member. Let S denote the set of numbers a with the property that a is smaller than all members of T. We know that $1 \in S$. Now if a is a member of S we cannot have $a + 1$ being a member of T, because a is less than all members of T and so $a + 1$ would be the least member of T. This argument shows that if $a \in S$ then $a + 1 \in S$. So the axiom of induction implies that $S = \mathbb{N}$. This would imply that T had no members at all. This is a contradiction, and so the assumption that T has no least member must be false.   ●

● Proposition 2

The well-ordering principle implies the axiom of induction.   ●

PROOF

Let S denote a subset of $\mathbb{N}$ having the two properties

(i) $1 \in S$,

(ii) if $k \in S$ then $\mathrm{succ}(k) \in S$.

We have to prove that the well-ordering principle implies that $S = \mathbb{N}$. Let T denote the set of numbers not in S. If T is non-empty then it has a least member, denoted by a. Property (i) implies that $a \neq 1$, so $a - 1 \in \mathbb{N}$ and $a - 1 \notin T$, and therefore $a - 1 \in S$. But now property (ii), with k replaced by $a - 1$, implies that $a \in S$ and so $a \notin T$. We have shown that $a \in T$ and $a \notin T$, which is a contradiction. This is an example of *reductio ad absurdum*, discussed in §1.3. We must therefore have T empty, i.e. $S = \mathbb{N}$.   ●

Notice that we have used subtraction in this proof, without in fact having defined it. In a complete account of the foundations of the number system we would do so after the development of addition as outlined above.

2.2 Proof by Mathematical Induction

The axiom of induction is the basis of an important method of proof known as 'Proof by Mathematical Induction'. This involves proving statements involving a variable n, typical examples being

(i) for all $n \in \mathbb{N}$, $1^3 + 2^3 + 3^3 + \ldots + n^3 = n^2(n + 1)^2/4$,

(ii) for all $n \geq 3$, $2^n \geq 1 + 2n$,

(iii) for all $n \geq 4$, $n! > 2^n$.

Each of these statements is of the form $P(n)$, and the variable n ranges over all natural numbers from some initial number onwards. To explain the procedure we shall take this initial number to be 1, without any loss of generality. The procedure is as follows. We first establish that $P(1)$ is true. Now let us suppose that n is the first number for which we do not (yet) know that $P(n)$ is true. (Another way of stating this is to suppose that $P(k)$ is true for $1 \le k \le n-1$.) From this supposition we aim to deduce that $P(n)$ is true. Before discussing this further we shall justify the procedure. Let S be the set of numbers n for which $P(n)$ is true. The procedure shows that $1 \in S$ and that if $n-1 \in S$ then $n \in S$. The axiom of induction then says that $S = \mathbb{N}$, i.e. that $P(n)$ is true for all $n \in \mathbb{N}$. In practice, it is often the case that $P(n)$ is deduced using only $P(n-1)$, as in examples 1 and 2 below. In some cases, like Example 3, we deduce $P(n)$ from $P(k)$ for some value(s) of k between 1 and $n-1$.

Both stages in the procedure of proof by induction are important. The second stage, referred to as the *inductive step*, is clearly needed, and is usually the major part of the proof. The supposition that $P(k)$ is true for $1 \le k \le n-1$, used in proving the implication, is called the *inductive hypothesis*. The first stage, showing that the statement is true for some starting value (in many cases $n = 1$), is equally important. (It is often referred to as the *anchor* or *initialization* for the proof.) For example, the implication $n-1 = n \Rightarrow n = n+1$ is true, even though the constituent equations are both invalid, because adding 1 to both sides of an equation is a valid algebraic operation. This corresponds to the inductive step in a proof. However, this does not mean that we can show that $n = n+1$ for all n. There is no anchor; no initial value of n from which a set of true statements can begin. We shall now consider some examples of this procedure. In the first we shall label the components of the proof explicitly. Later we drop that practice, but these stages are always present.

Some accounts of mathematical induction make a distinction between the case where the inductive hypothesis supposes that $P(k)$ is true for $1 \le k \le n-1$, as we have done, and the simpler form of the hypothesis which supposes only $P(n-1)$. The first situation is sometimes referred to as *strong induction*, but in fact the two forms are equivalent.

Example 1

Prove by mathematical induction that $\displaystyle\sum_{i=1}^{n} i^3 = \frac{n^2(n+1)^2}{4}$ for all $n \in \mathbb{N}$.

Notice that we are using the summation notation as explained above after Tutorial Problem 1, in this case with $f(i)$ replaced by i^3.

The anchor step

When $n = 1$ both sides of the equation are equal to 1, so the result is true.

The inductive hypothesis

Suppose the result is true for $1 \le k \le n-1$, i.e. that $\displaystyle\sum_{i=1}^{k} i^3 = \frac{k^2(k+1)^2}{4}$ for $1 \le k \le n-1$.

The inductive step

We shall show that $P(n-1) \Rightarrow P(n)$ where $P(n)$ stands for the equality we are trying to establish. We have

$$\sum_{i=1}^{n} i^3 = n^3 + \sum_{i=1}^{n-1} i^3 = n^3 + \frac{(n-1)^2 n^2}{4} \quad \text{(using the inductive hypothesis)}$$

$$= \frac{4n^3 + (n-1)^2 n^2}{4} = \frac{n^2(n^2 + 2n + 1)}{4} = \frac{n^2(n+1)^2}{4}.$$

Conclusion

We have shown that we can deduce $P(n)$ from the hypothesis that $P(n-1)$ is true, and so together with $P(1)$ this establishes by induction that the result is true for all values of n.

Firstly, we note the contraction 'by induction' is often used instead of 'by mathematical induction'. More importantly though, we should discuss the phrase 'suppose that the result is true for $1 \le k \le n - 1$'. This looks as if we are assuming what we want to prove, and this causes doubts in the minds of many concerning the validity of the method, despite the explanation in terms of the axiom of induction. To understand this we should recall that the procedure in Example 1 involved establishing the truth of the statement $P(n-1) \Rightarrow P(n)$. We discussed statements of implication in some detail in §1.3, and there we noted that a statement of the form 'p implies q' is making no assertion about the truth of p or of q. The truth of such an implication simply asserts the validity of the deduction which the implication contains. This is emphasized by the conditional 'if' in the version 'if $P(n-1)$ then $P(n)$'. The discussion in §1.3 made it clear that a statement of implication $p \Rightarrow q$ is true when the premise p is false, and so when we say 'suppose that $P(n-1)$ is true' this is to be regarded as the beginning of a demonstration of the truth of the implication $P(n-1) \Rightarrow P(n)$, where we have taken it for granted that if $P(n-1)$ were false then the implication would certainly be valid, in view of §1.3.

Given this discussion, the question will often be asked as to whether we can *prove* that the method of induction is valid. Some books attempt to do this by appealing to the well-ordering principle, which some find intuitively more appealing than induction. However, this would be circular, because we have shown that the axiom of induction and the well-ordering principle are in fact logically equivalent. So what we have to realize is that the validity of the method of proof by induction is at root a matter of an axiom. We *choose* to accept the method as valid, just as in fact we *choose* to accept as valid all other logical procedures we use. It may be of interest to note that not all mathematicians share this view. Some would regard the procedures of logic as existing *a priori* rather than being human inventions. This is a point of view often ascribed to the Ancient Greek philosopher Plato.

TUTORIAL PROBLEM 2

Discuss in your tutorial group the logical principles underlying the procedure of proof by mathematical induction. Discuss your reasons for belief in its validity.

In the next example the anchor for the proof of the first result is different from 1. The necessity for this is made clear in the statement of the problem.

Example 2

(a)   Prove by induction that $2^n \geq 1 + 2n$ for all $n \geq 3$.

(b)   Prove by induction that $n^2 + 3n$ is divisible by 2 for all $n \in \mathbb{N}$.

(a) When $n = 3$ the statement says $2^3 \geq 1 + 2 \times 3$, which is true. Suppose that $2^{n-1} \geq 1 + 2(n-1)$. We then have

$$2^n = 2 \times 2^{n-1} \geq 2(1 + 2(n-1)) \quad \text{(by the inductive hypothesis)}$$
$$= 2 + 4(n-1) = (2 + 2(n-1)) + 2(n-1) > 3 + 2(n-1) = 1 + 2n.$$

(We have used $2 + 2(n-1) > 3$ here. This is certainly true because $n - 1 \geq 3$.)

This has shown that if the inequality is valid for $n - 1$ then it is valid for n. This is the inductive step which, together with the anchor step, proves the inequality for all $n \geq 3$.

In fact, in this example the inequality is not valid for $n = 1$ or for $n = 2$, so that neither of these would have served as an anchor value.

(b) When $n = 1$, $n^2 + 3n = 4$, which is divisible by 2. This establishes the anchor for the induction.

Now suppose the result is true for $n - 1$. We then have

$$n^2 + 3n = (n-1)^2 + 2(n-1) + 1 + 3(n-1) + 3$$
$$= ((n-1)^2 + 3(n-1)) + 2(n+1).$$

The first term, $(n-1)^2 + 3(n-1)$, is divisible by 2 by the inductive hypothesis. The second term, $2(n+1)$, is divisible by 2 because of the factor of 2. So the whole expression is divisible by 2. We have therefore deduced the result for n, and so by induction the result is true for all $n \in \mathbb{N}$.

Example 3

Prove that for all $n \geq 2$, n can be expressed as a product of primes. (In the case of a prime number itself we allow the 'product' to contain just the one factor.)

By the convention above, 2 is a product of primes (with just the single factor 2). Now suppose that k can be expressed as a product of primes for $2 \leq k \leq n - 1$. Consider the number n. If it is a prime number then it is a (one-factor) product of primes. If it is not prime then it can be expressed as a product $n = r \times s$ where r and s are both numbers between 2 and $n - 1$. The inductive hypothesis asserts that each of these two numbers can be expressed as a product of primes, and this shows that n itself is also a product of primes. So by induction every $n \geq 2$ can be written as a product of prime numbers.

EXERCISES 2.2

1. Prove the following by the method of mathematical induction,

    (i) $\sum_{i=1}^{n}(2i-1)=n^2$,   (ii) $\sum_{r=1}^{n}r=\dfrac{n(n+1)}{2}$,   (iii) $\sum_{s=0}^{n}x^s=\dfrac{1-x^{n+1}}{1-x}$ $(x\neq 1)$.

2. Prove by induction that for $x > -1$, $(1+x)^n \geq 1+nx$ for all $n \in \mathbb{N}$.

3. Prove by induction that $5^n + 2.3^{n+1} + 1$ is divisible by 4 for all $n \in \mathbb{N}$.

4. Consider the inequality $n! \geq 5^n$. Find the smallest integer H for which this inequality is true, and then prove by induction that $n! \geq 5^n$ for all $n \geq H$.

5. A set of isosceles right-angled triangles is constructed as follows. The first triangle has sides containing the right-angle being of unit length. The second has as its hypotenuse one of the shorter sides of the first. The third has as its hypotenuse one of the shorter sides of the second, and so on. Fig 2.1 shows the construction as far as the fifth triangle.

    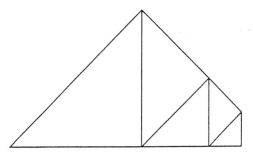

    Fig. 2.1 The first five triangles for Exercise 5.

    Find the length of the hypotenuse for each of the first five triangles. Write down a conjecture based on this pattern for the length of the hypotenuse of the nth triangle. Prove your result by induction. Find a formula for the total area of the first n triangles.

6. Let $S(n)$ denote the sum of the first n natural numbers, with alternating signs, for example $S(8) = 1 - 2 + 3 - 4 + 5 - 6 + 7 - 8$. Write down the first 10 values of $S(n)$ and use these to conjecture formulae for $S(2n)$ and $S(2n+1)$. Prove your conjectures by induction. Try to devise a single formula for $S(n)$ which is valid for both even and odd values of n.

2.3 Negative Integers

Peano's Axioms were formulated specifically to describe the natural numbers, or positive integers, and it can be shown that there is essentially only one system satisfying them, in the sense that any apparently different system has only superficial changes, such as the names or symbols used for the individual elements. One of the things we did not discuss was Peano's use of the axioms to prove the rules of algebra, things like $a + b = b + a$, $a \times 1 = a$, etc. Another approach to describing number systems is to list the rules of algebra as axioms. A list of such axioms will generally encompass many systems which differ from one another in structure as well as in surface detail. Such axioms are at the root of the subject of abstract algebra, with its various branches, such as group theory and the theory of rings and fields. In Examples 4, 5 and 6 below we have used some of these rules. We have given their names and also their algebraic formulations. These can all be proved by induction, and Peano did some of this in his 1889 booklet. We shall discuss these axioms more systematically in the context of rational numbers in Chapter 3. In this section we shall approach the negative integers by looking only at the additional algebraic rules which the system $\mathbb{Z}$ of all integers (positive, negative and zero) satisfies but which $\mathbb{N}$ does not. These are as follows.

(i) There is a number in $\mathbb{Z}$, denoted by 0, with the property that $a + 0 = a$ for all $a \in \mathbb{Z}$. The number 0 is referred to as the zero, or the additive identity.

(ii) For every number $a \in \mathbb{Z}$ there is a number $x \in \mathbb{Z}$ which satisfies the equation $a + x = 0$. The number x is called the additive inverse of a, and denoted by $-a$.

Notice that in (ii) the order of quantification, as discussed in §1.5, is important. Expressing (ii) in the symbolic form introduced in that section would give $\forall a \in \mathbb{Z}, \exists x \in \mathbb{Z}, a + x = 0$. In many situations in algebra where we try to extend a system, it is in order to be able to solve equations which have no solutions in the parent system. In this case we cannot, for example, solve the equation $3 + x = 0$ within $\mathbb{N}$, but we can solve it within $\mathbb{Z}$.

We shall now look briefly at the use of these additional rules of algebra to prove some of the familiar properties of negative numbers. As we go through these examples we shall highlight the underlying rules of algebra being used as they occur. The first example might suggest that we have gone completely daft! Readers can be assured however that mathematicians do not normally spend time repeatedly multiplying zero by itself!

Example 4

Prove that $0 \times 0 = 0$.

$0 + 0 = 0$	rule (i) with $a = 0$
$0 \times (0 + 0) = 0 \times 0$	multiplying both sides by 0
$(0 \times 0) + (0 \times 0) = 0 \times 0$	'distributive rule': $a \times (b + c) = (a \times b) + (a \times c)$
$((0 \times 0) + (0 \times 0)) + x = (0 \times 0) + x$	adding x to both sides, where x is the solution of $(0 \times 0) + x = 0$ guaranteed by rule (ii)
$(0 \times 0) + ((0 \times 0) + x) = (0 \times 0) + x$	'associative rule': $(a + b) + c = a + (b + c)$

$$(0 \times 0) + 0 = 0 \qquad \text{using the property of } x$$
$$(0 \times 0) = 0 \qquad \text{using rule (i) with } a = 0 \times 0$$

Example 5

Prove that $a \times 0 = 0$ for all $a \in \mathbb{Z}$.

$$a + 0 = a \qquad \text{rule (i)}$$
$$0 + a = a \qquad \text{'commutative rule': } a + b = b + a$$
$$a \times (0 + a) = a \times a \qquad \text{multiplying both sides by } a$$
$$a \times 0 + a \times a = a \times a \qquad \text{distributive rule}$$
$$(a \times 0 + a \times a) + x = a \times a + x \qquad \text{adding } x \text{ to both sides, where } x \text{ is the solution of}$$
$$\qquad (a \times a) + x = 0 \text{ guaranteed by rule (ii)}$$
$$a \times 0 + (a \times a + x) = a \times a + x \qquad \text{associative rule}$$
$$a \times 0 + 0 = 0 \qquad \text{using the property of } x$$
$$a \times 0 = 0 \qquad \text{using (i) with } a \text{ replaced by } (a \times 0)$$

Notice that in Example 5 we have used less brackets than in Example 4. We have assumed the usual convention of multiplication having precedence over addition (many will have learned a mnemonic such as **BODMAS** for the precedence conventions in algebra).

Example 6

Prove that $(-a) \times (-b) = a \times b$.

In algebra we normally write ab instead of $a \times b$, and we shall do that in this example.

$$(a + (-a))(b + (-b)) = 0 \times 0 = 0 \qquad \text{property (ii) and Example 5}$$

$$ab + (-a)b + a(-b) + (-a)(-b) = 0 \qquad \text{expansion (distributive rule)}$$

$$ab + (-a)b + a(-b) + (-a)(-b) + (-a)(-b) = (-a)(-b) \qquad \text{adding } (-a)(-b) \text{ to both sides}$$

$$ab + (-a)b + (-a)(-b) + a(-b) + (-a)(-b) = (-a)(-b) \qquad \text{rearranging (associative rule)}$$

$$ab + (-a)(b + (-b)) + (a + (-a))(-b) = (-a)(-b) \qquad \text{using the distributive rule}$$

$$ab + (-a) \times 0 + 0 \times (-b) = (-a)(-b) \qquad \text{using property (ii)}$$

$$ab + 0 + 0 = (-a)(-b) \qquad \text{Example 5 and the commutative rule}$$

$$ab = (-a)(-b) \qquad \text{using the property of } 0$$

TUTORIAL PROBLEM 3

Discuss in your tutorial group how you were introduced to negative numbers at school. Try to remember how the rule 'minus times minus equals plus' proved in Example 6 was explained at school. Try to devise an explanation of this rule which would be convincing within the context of school mathematics.

2.4 Division and the Highest Common Factor

Further exploration of the integers concerns arithmetic matters, things like prime numbers, integer solution of equations like $x^2 + y^2 = z^2$, and many other topics. This is the subject of Number Theory, which is not part of this book. We shall explore just one topic, that of the highest common factor (h.c.f.) of two numbers. We first prove the fundamental result concerning division. When we divide one integer into another we obtain a quotient and a remainder. Proving that this always happens illustrates both the use of the well-ordering principle and proof by contradiction.

● *Proposition 3*

Given integers a and $b > 0$, there are integers q and r which satisfy

$a = bq + r$ where $0 \leq r < b$.

Furthermore, q and r are unique. ●

PROOF

Let T denote the set of numbers of the form $a - bt$, where t can be any integer. If $a \geq 0$ then $t = 0$ gives $a - bt = a \geq 0$. If $a < 0$ then $t = a$ gives $a - bt = a(1 - b) \geq 0$. So T always contains non-negative numbers, and so it contains a smallest non-negative number, using the well-ordering principle. Let r denote this number, and let q be the corresponding value of t. Now if $r \geq b$ then since $r = a - bq$ we have $r - b \geq a - bq - b = a - b(q - 1) \geq 0$, so that r would not be the smallest non-negative number in T. This is a contradiction, and so $r \geq b$ must be false. Thus, $0 \leq r < b$.

We now have to prove uniqueness. This is typical of many such proofs, where we assume two possibilities and show that in fact they are identical. So we suppose that $a = bq_1 + r_1$ and $a = bq_2 + r_2$, where $0 \leq r_1 < b$ and $0 \leq r_2 < b$. Equating the two expressions for a gives $bq_1 + r_1 = bq_2 + r_2$, and so $b(q_1 - q_2) = r_2 - r_1$. Now the left-hand side of this equation is an integer multiple of b. But because r_1 and r_2 are both between 0 and b we must have $-b < r_2 - r_1 < b$. The only integer multiple of b strictly between $-b$ and b is zero. So both sides of the last equation must be zero, giving $q_1 = q_2$ and $r_1 = r_2$, proving uniqueness. ●

Note that if $r = 0$ then $a = bq$, i.e. if the remainder is zero then a is an integer multiple of b. Another way of expressing this is to say that b is a divisor (factor) of a, abbreviated to $b|a$. Given two integers a and b, the integer k is a common divisor of a and b if $k|a$ and $k|b$. A number $d > 0$ is a greatest common divisor (g.c.d.), or highest common factor (h.c.f.) of a and b if $d|a$ and $d|b$, and if $k|a$ and $k|b$ implies that $k|d$. Two questions immediately arise: how do we know that there is such a number, and is it unique? The next result establishes this.

● Proposition 4

Let a and b be two non-zero integers. The least positive integer of the form $ax + by$, where x and y are also integers, is the unique h.c.f. of a and b. ●

PROOF

Taking $x = a$ and $y = b$ shows that there are positive integers of the form $ax + by$. The well-ordering principle guarantees the existence of a smallest such positive integer, d. So $d = as + bt$ for some integer values of s and t. Now any common divisor of a and b is a divisor of $as + bt$ and therefore a divisor of d. We now have to prove that d itself is a divisor of both a and b. Suppose that d is not a divisor of a. Proposition 3 tells us that $a = qd + r$, where $0 < r < d$. (We cannot have $r = 0$ since otherwise $d|a$.) But

$$r = a - qd = a - q(as + bt) = a(1 - qs) + b(-qt).$$

So r is a positive integer of the form $ax + by$. This is a contradiction, because $r < d$ and d is the least such positive integer. So we must have $d|a$, and similarly $d|b$. To prove uniqueness, suppose that d_1 and d_2 are both a h.c.f. of a and b. The definition of h.c.f. tells us that since d_2 is a h.c.f., $d_2|d_1$. Similarly $d_1|d_2$ and so $d_1 = d_2$, since they are both positive integers. ●

The question now arises as to how we can find values of x and y, and hence find the h.c.f. This is achieved through the Euclidean Algorithm, which determines the h.c.f. first and then enables x and y to be calculated. The values of x and y are not unique. This algorithm appears in Euclid's *Elements*, Book VII, Proposition 2. The term used by Euclid is usually translated as 'greatest common measure', reflecting the idea that numbers were thought of as synonymous with measurement at the time (around 300 BC).

2.5 The Euclidean Algorithm

Suppose a and b are two positive integers, and assume without loss of generality that $a > b$, and that b is not a factor of a. We can therefore divide b into a and, from Proposition 3, obtain $a = bq + r$, with $r \neq 0$. It will be convenient to have names for numbers playing the roles a and b, and if we imagine b divided into a written in the notation of fractions (a/b) then it makes sense to call a the numerator and b the denominator, and we shall do this in the equation $a = bq + r$. The Euclidean Algorithm proceeds in stages as follows. At the first stage, the quotient q and the remainder r are denoted by q_1 and r_1 respectively, signifying that they are the first of a set of such numbers:

$$a = bq_1 + r_1, \qquad\qquad 0 < r_1 < b.$$

The denominator now becomes the numerator, and the remainder becomes the denominator.

$$b = r_1q_2 + r_2, \qquad\qquad 0 < r_2 < r_1$$

The process is now repeated over and over again.

$$r_1 = r_2q_3 + r_3, \qquad\qquad 0 < r_3 < r_2$$
$$\vdots$$
$$r_{m-2} = r_{m-1}q_m + r_m, \qquad\qquad 0 < r_m < r_{m-1}$$
$$r_{m-1} = r_mq_{m+1}.$$

From the inequalities for the remainders we can see that they form a decreasing set of non-negative integers. Such a process cannot therefore continue for ever, and there must come a stage at which the remainder is zero. This is signified by the last step shown above. The last remainder r_m turns out to be the h.c.f. of a and b. We can see this by tracing through the steps in the algorithm. The last step tells us that r_m is a divisor of r_{m-1}. Therefore, in the penultimate step, r_m is a divisor of both terms on the right-hand side, and so is a divisor of r_{m-2}. This argument is repeated step by step, eventually telling us that r_m is a divisor of b, and then of a.

We have shown that the last remainder r_m is a common divisor of a and b. We now need to show that any other common divisor is a factor of r_m, proving that this is the h.c.f. The first equation in the algorithm tells us that if $d|a$ and $d|b$ then $d|r_1$. From the second step we can now deduce that $d|r_2$. This argument proceeds through all the stages of the algorithm, showing that d is a divisor of all the remainders, and in particular of the last one, r_m.

When two numbers have no common factor, the algorithm gives 1 as the last remainder, as in Example 7 below. It is convenient to say that, in this case, the h.c.f. is 1, so that two numbers always have a h.c.f. Euclid treats this case separately, as Proposition 1 in Book VII of *Elements*. He says that two such numbers are prime to one another, and we use the adjective coprime in these circumstances. He describes the situation by saying that the two numbers have no common measure, and that the process ends with unity. It is difficult for us to appreciate the difference in conception which this distinction reflects, because we have a firm notion of numbers as abstract entities, used for a variety of purposes apart from measurement.

So far the algorithm has been shown in general terms, and so really to understand it we need some numerical examples. In the first example we illustrate the case of coprime integers, and we have also indicated the process of replacing numerator by denominator as described at the beginning of this section.

Example 7

Use the Euclidean Algorithm to find the h.c.f. of 8273 and 482.

$$8273 = 482 \times 17 + 79,$$

$$482 = 79 \times 6 + 8,$$

$$79 = 8 \times 9 + 7,$$

$$8 = 7 \times 1 + 1,$$

$$7 = 1 \times 7.$$

So the h.c.f. of 8273 and 482 is 1.

Because the same step is repeatedly used in the algorithm, it is easy to program for a calculator or a computer in any language which allows loops. For example, the following segment of Pascal would produce the steps of the algorithm line by line (with a proper formatting of the `writeln` statement), when incorporated in a complete program.

```
num:=a;
den:=b;
repeat
        quo:=num div den;
        rem:= num mod den;
        writeln("num = quo x den + rem");
        num:=den;
        den:=rem;
until rem=0;
```

A number of languages including Pascal and BBC Basic have operators `div` and `mod` to find the quotient and remainder respectively from an integer division. The last two indented statements achieve the interchange of numbers shown by dashed lines in the last example.

Example 8

Use the Euclidean Algorithm to find the h.c.f. of 596 and 328. Find all possible integers x and y so that $596x + 328y$ is equal to the h.c.f.

$$596 = 328 \times 1 + 268, \qquad (1)$$
$$328 = 268 \times 1 + 60, \qquad (2)$$
$$268 = 60 \times 4 + 28, \qquad (3)$$
$$60 = 28 \times 2 + 4, \qquad (4)$$
$$28 = 4 \times 7.$$

The last remainder is 4, and so the h.c.f. of 596 and 328 is 4. To find x and y we 'unscramble' the algorithm as follows.

$$4 = 60 - 2 \times 28 \qquad \qquad \text{from (4)}$$
$$= 60 - 2(268 - 4 \times 60) \qquad 28 = 268 - 4 \times 60 \text{ from (3)}$$
$$= -2 \times 268 + 9 \times 60$$
$$= -2 \times 268 + 9(328 - 1 \times 268) \qquad 60 = 328 - 1 \times 268 \text{ from (2)}$$
$$= 9 \times 328 - 11 \times 268$$
$$= 9 \times 328 - 11(596 - 1 \times 328) \qquad 268 = 596 - 1 \times 328 \text{ from (1)}.$$
$$\text{So } 4 = -11 \times 596 + 20 \times 328.$$

So $x = -11$ and $y = 20$ is one pair of possible values.

Now suppose that x and y is any other pair of values satisfying $4 = 596x + 328y$. Using the particular values found above tells us that

$$596 \times (-11) + 328 \times 20 = 596x + 328y$$
$$328(20 - y) = 596(11 + x).$$

Dividing both sides by 4 (the h.c.f.) gives

$$82(20 - y) = 149(11 + x).$$

Now 82 and 149 have no factors in common, and so $11 + x$ must be a multiple of 82. So let $11 + x = 82m$, giving $x = 82m - 11$. We now substitute to obtain $82(20 - y) = 149 \times 82m$, and so $20 - y = 149m$, giving $y = 20 - 149m$. We have therefore found all the solutions, and we can write

$$4 = 596(82m - 11) + 328(20 - 149m).$$

The method used here will work for any numerical example.

EXERCISES 2.5

1. Use the Euclidean Algorithm to find the h.c.f. for the following pairs of numbers a and b,

    (i) 87 and 72,

    (ii) 1073 and 145,

    (iii) 7537 and 8039.

    In each case find all the pairs of integers x and y for which $ax + by$ is equal to the h.c.f.

2. Let a and b be two integers, and let d denote the h.c.f. Prove that if $d|c$ then $ax + by = c$ has integer solutions for x and y.

Suppose now that d is not a divisor of c, so that we can write $c = dq + r$, where $0 < r < d$. Use this, together with the results of Proposition 4, to show that the assumption that $ax + by = c$ has integer solutions, leads to a contradiction.

(These two together show that a necessary and sufficient condition for $ax + by = c$ to have integer solutions is that the h.c.f. of a and b is a divisor of c.)

3. Use the result of Exercise 2 to determine in each case whether the following equations have integer solutions. Find all the solutions for those that do.

    (i) $301x + 84y = 5$,

    (ii) $345x + 735y = 60$,

    (iii) $87x + 53y = 13$.

4. Prove that if d is the h.c.f. of a and b then nd is the h.c.f. of na and nb.

5. Write a program, for your calculator or computer, which will accept two integers as input, exhibit the steps of the Euclidean Algorithm and finally tell you the h.c.f. of the two integers.

6. Let a and b denote two integers. A number k is a common multiple of a and b if $a|k$ and $b|k$. Show that the set of common multiples contains positive integers. Let m denote the least common multiple (l.c.m.), whose existence is guaranteed by the well-ordering principle. Prove that if k is any common multiple of a and b then $m|k$. [Assume the contrary and then use the result of Proposition 3 applied to m and k to derive a contradiction.]

2.6 Digital Representation

Our digital notation for integers is based on the number ten, in the sense that 7234 is short for $7 \times T^3 + 2 \times T^2 + 3 \times T + 4$, where T stands for the number ten—for most people equal to the number of their fingers (digits). The convention of using ten as the base for number notation is a very old one, but by no means universal, as many books on the history of mathematics relate. When we consider the notation with T used in place of ten as above, it is clear that there is no reason why T should not be replaced by other numbers. In common use in recent times, in connection with computers, are the bases 2 (binary), 8 (octal) and 16 (hexadecimal). In the decimal system the single digits are the familiar ones $0, 1, \ldots, 9$. In the binary system only 0 and 1 are used. In octal we have $0, 1, \ldots, 7$ and in hexadecimal we use $0, 1, \ldots, 9, A, B, C, D, E, F$ for single digits, so that, for example, D in hexadecimal is equivalent to 13 in decimal notation, 15 in octal and 1101 in binary. There are many interesting patterns which can be observed in various digital notations. For example, if numbers are expressed in base six then they are divisible by three if and only if the right-hand digit is either 3 or 0. Readers who are interested are invited to explore some of the patterns which can arise. To do this we need to be able to convert from one number base to another. Because we are so familiar with base ten we shall restrict ourselves here to conversions to and from base 10, and we shall illustrate the methods through examples.

Example 9

The number 7521 is expressed in base eight. Convert it to base ten.

As the number is in base eight, it represents $7 \times 8^3 + 5 \times 8^2 + 2 \times 8 + 1$. This gives 3921 in base ten.

Example 10

The number 2E5A is expressed in the hexadecimal system. Convert it to base ten.

In hexadecimal A and E represent 10 and 14 respectively in decimal. So in base ten 2E5A represents

$$2 \times 16^3 + 14 \times 16^2 + 5 \times 16 + 10 = 11866.$$

Example 11

The number 654 is expressed in base ten. Convert it to base seven.

We apply continued division by 7 as follows, using the quotient/remainder notation.

$$654 = 93 \times 7 + 3, \tag{1}$$
$$93 = 13 \times 7 + 2, \tag{2}$$
$$13 = 1 \times 7 + 6, \tag{3}$$
$$1 = 0 \times 7 + 1.$$

We stop when the quotient first becomes zero. The digits of the representation in base seven are then obtained by reading off the remainders in reverse order, giving 1623. The justification is given by reversing the process.

$$13 \times 7 = (1 \times 7 + 6) \times 7 \qquad\qquad \text{from (3)},$$
$$= 1 \times 7^2 + 6 \times 7.$$
$$93 = 1 \times 7^2 + 6 \times 7 + 2 \qquad\qquad \text{substituting in (2)}.$$
$$93 \times 7 = (1 \times 7^2 + 6 \times 7 + 2) \times 7$$
$$= 1 \times 7^3 + 6 \times 7^2 + 2 \times 7.$$
$$654 = 1 \times 7^3 + 6 \times 7^2 + 2 \times 7 + 3 \qquad\qquad \text{from (1)}.$$

Example 12

Write a computer program to convert the positive integer n from base ten to base T.

Recall that if we have two numbers then in some languages, including Pascal and BBC Basic, mod gives the remainder and div gives the quotient when the two numbers are divided. The following segment of Pascal code implements the division procedure which we used in Example 11.

```
   b:=n;
   repeat
           a:=b mod T;
           write(a);
           b:=b div T;
   until b=0;
```

This can be incorporated in a complete program which prompts the user to input the number *n* and the base *T*. We are converting from base ten and so the assumption is that the computer reads numbers input in base ten. Note that the `write` statement will output the digits in reverse order, as they occurred in Example 11. Those readers who are proficient programmers can work out a way of making the output digits appear in the correct order.

As well as there being interesting patterns in this topic, it is also instructive to explore the algorithms of arithmetic in bases other than ten. A detailed exploration of this is unfortunately outside the scope of this book.

TUTORIAL PROBLEM 4

Try to devise a method of working out 153×245, where these numbers are written in base six, where the procedure must not involve conversion to base ten, and where all the arithmetic must be done in base six.

EXERCISES 2.6

1. Convert the date, month and year of your birth into binary numbers.

2. Complete a program along the lines of Example 12. Try to arrange the output with the digits in the correct order.

3. Devise a quick way of converting numbers from base four to hexadecimal.

4. Try to find a way of converting 24351 in base eight into base six, without an intermediate conversion to base ten.

5. Write a program to convert a number from base *S* to base *T*, where *S* and *T* are both less than ten. Discuss the problems involved when *S* or *T* is greater than ten.

6. Show that 144 is a square in base seven. What is it the square of? Is 144 a square in any other base (greater than four)? Explain your answer.

Summary

In this chapter we have looked at some properties of the set of integers. We have given just a taste of a number of different aspects, all of which are capable of much more detailed development. In looking at the foundations through the Peano

Axioms we have chosen a historically important contribution, which looked at the integers from a formally logical point of view. The method of proof by mathematical induction grows out of the axioms, and is important throughout mathematics. Readers who have not met this before are urged to master both the method and the underlying principles. In considering the negative integers, we have chosen to give just a glimpse of the methods used in abstract algebra, where properties have to be derived ultimately from a set of axioms which specify how the operations of the algebraic system work. We have covered just one topic concerned with the arithmetic of integers, which is at the beginning of the vast subject of number theory. The Euclidean Algorithm has algebraic applications as well as purely arithmetic ones, and again is historically of interest. Finally, we have looked very briefly at the mathematical principle underlying the way we write numbers, which is at the root of the familiar algorithmic processes of addition, subtraction, multiplication and division. In present day mathematics there is a good deal of emphasis on algorithms, and their implementation on computers.

EXERCISES ON CHAPTER 2

1. Prove the following by induction

   (i) $\sum_{j=1}^{n} j^2 = \dfrac{n(n+1)(2n+1)}{6}$, (ii) $\sum_{m=1}^{n} m2^{m-1} = (n-1)2^n + 1$.

2. Prove by induction that $10^n \geq n^{10}$ for all $n \geq 10$.

3. The binomial coefficients are defined as follows. Firstly we define $\binom{n}{0} = \binom{n}{n} = 1$ for all $n \geq 1$. For $0 < r \leq n$ we then define inductively $\binom{n+1}{r} = \binom{n}{r-1} + \binom{n}{r}$. Prove by induction that $\binom{n}{r} = \dfrac{n!}{(n-r)!r!}$.

4. Adapt the proof of Proposition 4 to show that if a, b and c are non-zero integers, then the least positive integer of the form $ax + by + cz$, where x, y and z are also integers, is the highest common factor of a, b and c.

5. Let $d = \text{h.c.f.}(a, b)$. Prove that $\text{h.c.f.}(a, b, c) = \text{h.c.f.}(d, c)$. Use this result to find the highest common factor of 682, 651, 527 using the Euclidean Algorithm.

6. Write a computer program to calculate the highest common factor of three integers using the Euclidean Algorithm.

3 • The Rational Numbers

We can learn a great deal about the way we think about numbers today by taking a historical perspective, and we shall do this very briefly before embarking on a study of the rational numbers. We begin, as with so many mathematical ideas, with the Ancient Greeks and, in particular with Pythagoras and his followers, who lived around 500 BC. The school of philosophy that they founded studied four subjects in what we might loosely call the scientific domain. These were arithmetic, geometry, music and astronomy. A basic tenet of the Pythagoreans was that number is everything, and this referred to positive whole numbers. This is not the place to describe their numerical approaches to music and astronomy, save to say that the phrase 'the music of the spheres' has its roots there. Pythagoras' Theorem, which we regard today as belonging to geometry, also has an arithmetical side, relating to such phenomena as the 3,4,5 triangle, which had been used in much earlier times as a device for measuring right-angles. Many solutions of the integer equation $a^2 + b^2 = c^2$ appear to have been known, and it may have been thought that all right-angled triangles fitted into such a pattern. This would have meant that given a right-angled triangle, one could find a sufficiently small unit of length so that each side of the triangle would be an integer multiple of this length, i.e. that each side could be 'measured' by this length. Pairs of lengths for which this could be done were said to be commensurable. The discovery, attributed to Pythagoras himself, that this was not the case is described in historical accounts as a 'crisis' in Greek mathematics. The fact that for an isosceles right-angled triangle the hypotenuse and another side were incommensurable cast doubt upon the whole basis of the Pythagorean view that everything was based on whole numbers. In modern terms it is the equivalent of saying that $\sqrt{2}$ is irrational. We discuss this further in Chapter 5.

Another point of view that is worth discussing is the relationship between number and measurement. Much of the geometry of Euclid concerning comparisons of lengths or areas would nowadays be expressed through numbers, for we identify number and length very closely. For Euclid however, his writing suggests that this was not the case. Many of his results are obtained by direct comparison of segments. His proof of the Euclidean Algorithm talks about one segment 'measuring' another. In our terms we would say that the length of one segment is an integer multiple of the length of another.

Proportionality was expressed in terms of relationships between whole numbers. Fractions were not regarded as single entities in the way in which we think of a half, for example, as an individual number. 'A half' seems to have been thought of as a relationship which reflected the fact that one number was twice another. In fact, this notion is at the root of the way we shall approach the construction of the rational number system in §3.2. Relationships of proportionality between incommensurable numbers caused particular difficulties. Some of the theory is attributed to the Greek mathematician Eudoxus, and it is his work which is generally considered to be the content of those parts of Euclid's *Elements* which deal with the subject.

Euclid is commonly said to have had no algebra. This is true only insofar as there was no algebra of numbers in the form we have it today. Some of the axioms we shall study in §3.1, which concern the rules of algebra, are in fact present in Euclid. The important difference is that they refer to a kind of geometrical algebra, which expresses relationships between line segments, areas and volumes. For example, Book II, Proposition 1 of Euclid's *Elements* states, 'If there be two straight lines, one of which is divided into any number of parts, the rectangle contained by the two straight lines is equal to the rectangles contained by the undivided line, and the several parts of the divided line'. (This translation is taken from an 1862 edition for schools.) In terms of numerical algebra this is equivalent to a statement of the distributive rule $a(b + c + d + \ldots) = ab + ac + ad + \ldots$. Euclid's geometrical algebra is highly developed, as anyone who studies the geometrical books of the *Elements* will agree.

It needs a deep and detailed study of Greek mathematics to be able really to appreciate the difference between their ideas and ours, but at least we should realize that today's view of number is the product of historical evolution, and may therefore change in the future. Our contemporary notion of number has therefore developed over a period of at least 3000 years. Our present conceptions of the real number system were formulated only about 100 years ago, and we shall discuss that in Chapter 5.

There are two points of view we can take in moving from a discussion of the integers to a consideration of the rational number system. One is to add extra axioms, as we did when we discussed the negative integers in §2.3. The second is to construct the rationals from the integers, through an abstract mathematical approach. Both of these will, of course, reflect our long-standing acquaintance with rational numbers and fractions from everyday mathematics and from school.

3.1 Solving Equations

To solve linear equations in the system of integers we need to be able to subtract, and §2.3 discussed procedures for defining subtraction. Within the system of rational numbers we need to be able to progress from multiplication to division (apart from division by zero which is inadmissible). Like subtraction, this can be done by specifying some algebraic axioms that the rational numbers must obey. In §2.3 we simply introduced the necessary extra axioms. In this section it is appropriate to give a comprehensive list for reference. Some axioms were referred to in Examples 4, 5 and 6 of Chapter 2. The axioms concern the algebraic operations of addition and multiplication. Besides these, there are axioms for ordering which deal with inequality and the way it interacts with addition and multiplication. The topic of inequality will be dealt with in Chapter 4 and so here we restrict discussion to the *algebraic* axioms.

In the rational number system $\mathbb{Q}$ there are two operations, addition, denoted by $+$, and multiplication, denoted by $\times$, which satisfy the following axioms

A1 (Closure) $\forall a, b \in \mathbb{Q}, a + b \in \mathbb{Q}$.

A2 (Associativity) $\forall a, b, c \in \mathbb{Q}, a + (b + c) = (a + b) + c$.

A3 (Additive Identity) $\exists z \in \mathbb{Q}, \forall a \in \mathbb{Q}, a + z = a$.

A4 (Additive Inverse) $\forall a \in \mathbb{Q}, \exists x \in \mathbb{Q}, a + x = z$.

A5 (Commutativity) $\forall a, b \in \mathbb{Q}, a + b = b + a$.

M1 (Closure) $\forall a, b \in \mathbb{Q}, a \times b \in \mathbb{Q}$.

M2 (Associativity) $\forall a, b, c \in \mathbb{Q}, a \times (b \times c) = (a \times b) \times c$.

M3 (Multiplicative Identity) $\exists e \in \mathbb{Q}, \forall a \in \mathbb{Q}, a \times e = a$.

M4 (Multiplicative Inverse) $\forall a \in \mathbb{Q}, a \neq z \Rightarrow \exists y \in \mathbb{Q}, a \times y = e$.

M5 (Commutativity) $\forall a, b \in \mathbb{Q}, a \times b = b \times a$.

D (Distributivity) $\forall a, b, c \in \mathbb{Q}, a \times (b + c) = (a \times b) + (a \times c)$.

This is a very abstract presentation, but is recorded here in this form because it is a set of axioms that is satisfied not just by $\mathbb{Q}$ but by a variety of other systems. Readers should try to put some of these statements into words. For example, A1 can be read as 'the sum of two rational numbers is another rational number'. Notice that A1 allows us to combine only two numbers, so to add more than two we have to combine them successively in pairs. Axiom A2 tells us that it does not matter which order we do this in; adding a to the sum of b and c gives the same result as adding c to the sum of a and b. Adding a fourth number d can again be done in any order, and so on. The same considerations apply to multiplication, as in axioms M1 and M2.

Those readers who have met modular arithmetic will find that all those systems satisfy the axioms when the modulus is a prime number, and that only M4 fails for composite moduli. Any such system is called a field, and these are studied in many contexts. In the case of the rational number field, the element z postulated in axiom A3 is of course zero. In M3, e is the number 1. However, in fields whose elements are not numbers it is important to use symbols which make it clear, for example, that in A3 we are not necessarily referring to the number 0. In A4 the number x is nothing other that what we define to be $-a$, as in 2.3. In M4 the number y is the reciprocal of a, denoted by $1/a$ or a^{-1}. Notice in M4 the condition $a \neq z$. This corresponds to the usual restriction 'you can't divide by 0'. In Example 5 in Chapter 2 we proved that in $\mathbb{Z}$ multiplying by zero always gives zero. The reasoning given there used axioms from the list above, and so the result is true in $\mathbb{Q}$ as well. This means that if $a = 0$ then $a \times y = 0$ for all values of y, so that there could not be a solution of $a \times y = 1$.

TUTORIAL PROBLEM I

> The discussion above, showing that division by zero is impossible, assumes that the numbers 0 and 1 (z and e in the axioms) are different. Discuss whether this follows from the axioms given, or whether it needs to be added to the list.

We shall not be using such an abstract approach very much, but those who are interested will find some examples within the exercises for this section.

These axioms are those which enable us to solve linear equations. They are all satisfied by the real and complex numbers as well as the rational numbers, and so the following example applies to all three number systems.

Example I

Show how the axioms for a field give the solution of the equation $ax + b = c$, where $a \neq 0$. Notice that we are following the usual algebraic convention of denoting multiplication by juxtaposition, i.e. we write ax instead of $a \times x$.

$$ax + b = c,$$

$(ax + b) + (-b) = c + (-b),$	using A4, which ensures the existence of $-b$,
$ax + (b + (-b)) = c + (-b),$	using A2,
$ax + 0 = c + (-b),$	using A4,
$ax = c + (-b),$	using A3,
$ax = c - b,$	using $c - b$ as an abbreviation for $c + (-b)$,
$a^{-1}(ax) = a^{-1}(c - b)$	using M4, which ensures the existence of a^{-1},
$(a^{-1}a)x = a^{-1}(c - b)$	using M2,
$(aa^{-1})x = a^{-1}(c - b)$	using M5,
$1 \times x = a^{-1}(c - b)$	using M4,
$x \times 1 = a^{-1}(c - b)$	using M5,
$x = a^{-1}(c - b)$	using M3.

● Definition I

We define the fraction $\dfrac{p}{q}$, where $p \in \mathbb{Z}$, $q \in \mathbb{Z}$ and $q \neq 0$, to be the solution x of the equation $qx = p$.

Example 2

Show how to use Definition 1 to add two fractions.

Let x and y be two fractions, where $qx = p$ and $sy = r$. Multiplying the first by s and the second by q gives $qsx = ps$ and $qsy = qr$. Adding these two equations and factorizing (axiom D) gives $qs(x + y) = ps + qr$. This demonstrates, from

Definition 1, that

$$\frac{p}{q} + \frac{r}{s} = \frac{ps + qr}{qs}.$$

TUTORIAL PROBLEM 2

Try to remember how you were introduced to multiplication and division of fractions at school. Try to formulate a convincing explanation of the rule 'turn upside down and multiply' for division of fractions. You may wish to tackle Exercises 1 and 2 first.

EXERCISES 3.1

1. Suppose that $px = q$ and that $qy = p$ $(pq \neq 0)$. Prove from the axioms that $xy = 1$, i.e. that y is the multiplicative inverse of x.

2. Let x and y be two fractions, where $qx = p$ and $sy = r$ $(p, q, r, s \in \mathbb{Z}$, $q \neq 0, s \neq 0)$. Use Definition 1 to obtain the usual expressions for $x - y$, xy and x/y (defined as xy^{-1}) using a similar approach to that of Example 2.

3. Prove that the additive identity in a field is unique, i.e. that if z_1 and z_2 both satisfy axiom A3 then $z_1 = z_2$. State explicitly which axioms you use in the course of your proof. Prove also that the multiplicative identity (axiom M3) is unique.

4. Prove, without using axiom M5, that $(a \times b)^{-1} = b^{-1} \times a^{-1}$ by showing that if $y = b^{-1} \times a^{-1}$ then $(a \times b) \times y = e$, thus demonstrating that $b^{-1} \times a^{-1}$ satisfies the definition of multiplicative inverse given through axiom M4.

3.2 Constructing the Rational Numbers

Deduction from axioms as a way of establishing the foundations of number systems is discussed in several places in this book, in particular in the previous section. In this chapter we shall therefore pay rather more attention to mathematical construction as a method of embedding the integers in the larger system of rational numbers. The underlying idea is that of fractions, which we are familiar with from school mathematics. The essential components of a fraction we have to work with are the numerator and denominator, both of which are integers. The fact that we conventionally write them with one over the top of the other, separated by a short horizontal line, is less important. We could write them in any way we like, so long as we knew which integer was the numerator and which the denominator. We make use of this idea to write fractions as ordered pairs of numbers, rather like coordinates. For example, instead of writing $\frac{1}{2}$ we could write $(1, 2)$ where the first component is the numerator and the second component is the denominator. The idea of an ordered pair like $(1, 2)$ is not contained within the fundamental language and symbolism of sets introduced in Chapter 1, where we remarked that in the specification of sets the order of the members of a set was immaterial. In fact, ordered pairs can be described in terms of sets in a way which was devised by the 20th century Polish mathematician Kuratowski.

● Definition 2

Given x and y (not necessarily distinct) the ordered pair (x, y) is defined by

$$(x, y) = \{x, \{x, y\}\}.$$ ●

TUTORIAL PROBLEM 3

Discuss Definition 2. In particular show that if $x \neq y$ then $\{x, \{x, y\}\} \neq \{y, \{y, x\}\}$, so that (x, y) and (y, x) are different. Show also that if $(x, y) = (y, x)$ then $x = y$. Try to extend these ideas to give a definition of an ordered triple (x, y, z) and prove some similar results concerning equality of triples.

Having established the notion of an ordered pair we are now in a position to formulate the construction of $\mathbb{Q}$.

● Definition 3

A fraction is defined to be an ordered pair (a, b) where a and b are integers, and $b \neq 0$.   ●

The purpose of this definition is to establish a logical connection back to the language of sets. We shall continue to use the traditional notation for fractions.

Now, a number like 'three-quarters' has many representations as fractions. We regard the following as fractions all of which represent the same number

$$\frac{3}{4}, \frac{6}{8}, \frac{15}{20}, \frac{300}{400}, \frac{-12}{-16}, \frac{3 \times 10^6}{4 \times 10^6}, \frac{-3}{-4}.$$

When we learn about this at school, we use the term 'equivalent fractions' to signify the relationship between them. This suggests that there is an equivalence relation, in the sense of §1.6, where we would expect the equivalence classes to be just the sets of equivalent fractions. Because we are trying to use the integers as the basis for our construction we must use only integer arithmetic in the definitions. The clue here comes from noticing that if we take any two of the fractions from the above list and cross multiply we obtain equal integers. For example with the first two in the list it is the case that $3 \times 8 = 4 \times 6$. This is a statement about integers.

● Definition 4

Two fractions $\dfrac{a}{b}$ and $\dfrac{c}{d}$ are said to be equivalent if $ad = bc$.   ●

● Proposition I

The relation described in Definition 4 is an equivalence relation.   ●

PROOF

We shall use the notation $\equiv$ for equivalence, as in §1.6.

(i) Reflexivity. For all $a, b \in \mathbb{Z}$, $ab = ba$ and so $\dfrac{a}{b} \equiv \dfrac{a}{b}$.

(ii) Symmetry. Suppose $\dfrac{a}{b} \equiv \dfrac{c}{d}$. Then $ad = bc$, and so, by commutativity for the integers, $cb = da$, showing that $\dfrac{c}{d} \equiv \dfrac{a}{b}$.

(iii) Transitivity. Suppose $\dfrac{a}{b} \equiv \dfrac{c}{d}$ and $\dfrac{c}{d} \equiv \dfrac{e}{f}$. Then $ad = bc$ and $cf = de$. Multiplying both left-hand sides and both right-hand sides together gives $adcf = bcde$. This factorizes to give $(af - be)cd = 0$. Now b, d, f are non-zero, being denominators. Since $ad = bc$ and $cf = de$ we conclude that if $c = 0$ then $a = 0$ and $e = 0$, giving $af = be$. If $c \neq 0$ then since $d \neq 0$, $(af - be)cd = 0$ implies that $af - be = 0$, i.e. $af = be$. This shows that $\dfrac{a}{b} \equiv \dfrac{e}{f}$, establishing transitivity. $\bullet$

Readers studying this proof may have wondered why, after having obtained the equation $(af - be)cd = 0$, we did not simply divide by cd in the case $c \neq 0$. Division is something which occurs within $\mathbb{Q}$ but not within $\mathbb{Z}$, and we are trying to construct $\mathbb{Q}$ using only the properties of the integers. We have, in fact, used the property that $pq = 0$ implies $p = 0$ or $q = 0$. This can be proved within the algebra of integers, and therefore without recourse to division.

We are now in a position to define rational numbers.

• Definition 5

The rational number system $\mathbb{Q}$ is the set of all equivalence classes corresponding to the equivalence relation given in Definition 4. $\bullet$

This is a very abstract definition, and we usually work with a representative from an equivalence class, so that we would take $\frac{1}{2}$ as a representative for the rational number 'one half'.

TUTORIAL PROBLEM 4

Try to show that each equivalence class contains a pair (a, b) for which the h.c.f. of a and b is 1. Show that there is a unique such pair for which b is positive. Finally, show that the equivalence class consists of the set $\{(ma, mb) : m \in \mathbb{Z}, m \neq 0\}$.

So far we have defined just the numbers themselves, and so the next task is to describe how to do arithmetic. As in the previous section we shall only discuss addition, leaving the details for the other arithmetic operations to the exercises.

There are some important principles here, relating to the fact that we have defined rational numbers as equivalence classes of fractions. Our task is not simply to state how to add fractions, but how to define addition of the equivalence classes. The method is to take a representative fraction from two equivalence classes, add them, and then define the sum of the classes to be the class containing the sum of the two fractions. There is a problem with this, namely that we have to be certain that if we were to choose different fractions we would always end up in the same equivalence class. The next result shows that this is indeed the case.

● *Proposition 2*

Suppose we have two rational numbers x and y and that we are given two representative fractions from each, so that

$$\frac{a_1}{b_1} \equiv \frac{a_2}{b_2} \quad \text{and} \quad \frac{c_1}{d_1} \equiv \frac{c_2}{d_2}.$$

We then have

$$\frac{a_1 d_1 + b_1 c_1}{b_1 d_1} \equiv \frac{a_2 d_2 + b_2 c_2}{b_2 d_2}. \qquad \bullet$$

PROOF
The equivalence of the given fractions tells us that $a_1 b_2 = a_2 b_1$ and that $c_1 d_2 = c_2 d_1$. We multiply the first equation by $d_1 d_2$ and the second by $b_1 b_2$. We then add them and factorize. This procedure gives

$$a_1 b_2 d_1 d_2 = a_2 b_1 d_1 d_2$$
$$c_1 d_2 b_1 b_2 = c_2 d_1 b_1 b_2$$
$$a_1 b_2 d_1 d_2 + c_1 d_2 b_1 b_2 = a_2 b_1 d_1 d_2 + c_2 d_1 b_1 b_2$$
$$(a_1 d_1 + b_1 c_1) b_2 d_2 = (a_2 d_2 + b_2 c_2) b_1 d_1.$$

From Definition 4 this equation tells us that

$$\frac{a_1 d_1 + b_1 c_1}{b_1 d_1} \equiv \frac{a_2 d_2 + b_2 c_2}{b_2 d_2}. \qquad \bullet$$

This kind of situation occurs in many places in algebra when we wish to combine equivalence classes in some way. The phraseology commonly used is to say, when we have proved such a result, that the operation (in this case addition) is 'well-defined'.

We introduced this section by saying that the integers could be embedded in the rationals. This means that there is a subset of the rational numbers which behaves in every way, both arithmetically and algebraically, like the integers. In the language of abstract algebra this subset would be said to be *isomorphic* to the integers. The subset in question comprises the equivalence classes containing fractions of the form $(n, 1)$, where $n \in \mathbb{Z}$. In Example 2 we derived, using Definition 1, the expression for the sum of two fractions, and we proved in Proposition 2 that this was compatible with the equivalence relation. This tells us that $(n_1, 1) + (n_2, 1) = (n_1 + n_2, 1)$, and so this subset is behaving exactly like the integers. Note that $n = 0$ is permitted, it is only the denominator which cannot be zero.

EXERCISES 3.2

1. Prove versions of Proposition 2 for the operations of subtraction, multiplication and division, as follows. Suppose we have two rational numbers x and y and that we are given two representative fractions from each, so that

$$\frac{a_1}{b_1} \equiv \frac{a_2}{b_2} \quad \text{and} \quad \frac{c_1}{d_1} \equiv \frac{c_2}{d_2}.$$

Prove that

$$\frac{a_1 d_1 - b_1 c_1}{b_1 d_1} \equiv \frac{a_2 d_2 - b_2 c_2}{b_2 d_2}, \quad \frac{a_1 c_1}{b_1 d_1} \equiv \frac{a_2 c_2}{b_2 d_2}, \quad \text{and} \quad \frac{a_1 d_1}{b_1 c_1} \equiv \frac{a_2 d_2}{b_2 c_2}.$$

3.3 Continued Fractions

Most readers will be familiar with the decimal expansion for fractions, for example

$$\frac{1}{10} = 0.1, \quad \frac{1}{8} = 0.125, \quad 2\frac{3}{20} = 2.15, \quad \frac{1}{3} = 0.333\ldots.$$

In this section we shall look briefly at a different expansion for rational numbers. Like decimals this can be extended to all real numbers. We find that continued fractions do not lend themselves to arithmetic procedures in the same way as decimals. One of their chief properties is that they give very good approximations with fewer terms than decimals.

There are references to continued fractions in the mathematics of the Arab world and in Indian writings at a time when European scientific culture was virtually non-existent, in the centuries around 500 AD. A modern algebraic theory can be traced back to the Renaissance Italian mathematician Bombelli, and thereafter through the work of Brouncker and Wallis in England, and Huygens, Lambert, Lagrange and others in continental Europe. In particular, the book *De Fractionibus Continius* by the great Swiss mathematician Leonhard Euler (1707–1783) contains much of what we use today of the algebraic theory of continued fractions. Like so much mathematics however, we can trace this topic back to Euclid, and we shall begin by recalling the calculations involved in the Euclidean Algorithm illustrated by Example 8 in Chapter 2.

$$596 = 328 \times 1 + 268,$$
$$328 = 268 \times 1 + 60,$$
$$268 = 60 \times 4 + 28,$$
$$60 = 28 \times 2 + 4,$$
$$28 = 4 \times 7.$$

We rewrite these calculations as

$$\frac{596}{328} = 1 + \frac{268}{328},$$
$$\frac{328}{268} = 1 + \frac{60}{268},$$
$$\frac{268}{60} = 4 + \frac{28}{60},$$
$$\frac{60}{28} = 2 + \frac{4}{28},$$
$$\frac{28}{4} = 7.$$

If we invert the second equation and substitute in the first we obtain

$$\frac{596}{328} = 1 + \frac{1}{\dfrac{328}{268}} = 1 + \frac{1}{1 + \dfrac{60}{268}}.$$

Substituting for $60/268$ by inverting the third equation then gives

$$\frac{596}{268} = 1 + \frac{1}{1 + \dfrac{1}{4 + \dfrac{28}{60}}}.$$

We can continue this process of inversion and substitution down to the end of the algorithm to obtain

$$\frac{596}{268} = 1 + \frac{1}{1 + \dfrac{1}{4 + \dfrac{1}{2 + \dfrac{1}{7}}}}.$$

We can see that the quotients from each stage of the algorithm appear as the denominators in the continued fraction. They are referred to as *partial quotients*. This means that we do not need to rearrange the steps of the algorithm by division as we have done here. We can simply read off the partial quotients from the algorithm itself. So, referring to Example 7 in Chapter 2 we can immediately write down

$$\frac{8273}{482} = 17 + \frac{1}{6 + \dfrac{1}{9 + \dfrac{1}{1 + \dfrac{1}{7}}}}.$$

This notation is very cumbersome, and various abbreviated notations have been devised. The two most common are illustrated as follows

$$\frac{596}{328} = 1 + \frac{1}{1+} \frac{1}{4+} \frac{1}{2+} \frac{1}{7}$$

$$\frac{8273}{482} = [17; 6, 9, 1, 7]$$

where the semicolon indicates the integer part as the first entry.

We have seen how to convert a rational number into a continued fraction. To perform the reverse process we should proceed from the right-hand end. Using the first example this would give

$$\frac{596}{328} = 1 + \frac{1}{1+} \frac{1}{4+} \frac{1}{2+} \frac{1}{7} = 1 + \frac{1}{1+} \frac{1}{4+} \frac{1}{2\frac{1}{7}}$$

$$= 1 + \frac{1}{1+} \frac{1}{4+} \frac{1}{15/7} = 1 + \frac{1}{1+} \frac{1}{4+} \frac{7}{15}$$

$$= 1 + \frac{1}{1+} \frac{15}{67} = 1 + \frac{67}{82} = \frac{149}{82}.$$

There are two things to notice about this. Firstly the procedure is very cumbersome. In fact, in order to obtain approximations we need to stop the continued fraction at various stages and convert each to a rational number. What we therefore need is a more systematic method, and we shall obtain a process beginning from the left instead of the right. Secondly, we notice that we do not end up with the fraction we started with. We found in Chapter 2 that the h.c.f. of 596 and 328 is 4, and if we cancel this highest common factor we obtain the fraction in its lowest terms, namely 149/82. This is always the case, and we shall see why.

In order to make progress we need to move from arithmetic examples to algebraic procedures. Suppose we have a continued fraction with successive partial quotients $a_1, a_2, a_3, \ldots$. If we stop this continued fraction at the nth partial quotient we obtain

$$c_n = a_1 + \frac{1}{a_2+} \frac{1}{a_3+} \cdots \frac{1}{a_n}.$$

This is called the nth *convergent* of the continued fraction. If we work out the first few of these by unscrambling the algebra in the same way that we converted the arithmetic example above to a fraction, an algebraic fraction will result. We will use p_n and q_n to denote the numerator and denominator respectively of the algebraic fraction arising from c_n. We find that

$$c_1 = a_1 = \frac{a_1}{1} = \frac{p_1}{q_1},$$

$$c_2 = a_1 + \frac{1}{a_2} = \frac{a_1 a_2 + 1}{a_2} = \frac{p_2}{q_2},$$

$$c_3 = a_1 + \frac{1}{a_2+} \frac{1}{a_3} = \frac{a_1 a_2 a_3 + a_1 + a_3}{a_2 a_3 + 1} = \frac{p_3}{q_3},$$

$$c_4 = a_1 + \frac{1}{a_2+} \frac{1}{a_3+} \frac{1}{a_4} = \frac{a_1 a_2 a_3 a_4 + a_1 a_2 + a_1 a_4 + a_3 a_4 + 1}{a_2 a_3 a_4 + a_2 + a_4} = \frac{p_4}{q_4}.$$

You are invited to verify the algebra involved in establishing these relationships.

At this stage we should begin to look for some relationships amid all this algebra, and if we look at the effect of the introduction of a_4 into c_4 we notice that we can write

$$c_4 = \frac{a_4(a_1a_2a_3 + a_1 + a_3) + (a_1a_2 + 1)}{a_4(a_2a_3 + 1) + a_2} = \frac{a_4p_3 + p_2}{a_4q_3 + q_2}.$$

Having noticed this relationship, we can look back to the previous case and see that

$$c_3 = \frac{a_3(a_1a_2 + 1) + a_1}{a_3(a_2) + 1} = \frac{a_3p_2 + p_1}{a_3q_2 + q_1}.$$

We will now prove that relationships between successive convergents always follow this pattern. It provides a further example of proof by induction.

● Proposition 3

Given the continued fraction

$$a_1 + \frac{1}{a_2+} \frac{1}{a_3+} \cdots \frac{1}{a_n+} \cdots,$$

if $c_n = p_n/q_n$ denotes the nth convergent, then for all $n \in \mathbb{N}$ we have the recurrence relations

$$p_n = a_np_{n-1} + p_{n-2},$$
$$q_n = a_nq_{n-1} + q_{n-2},$$

where we define $p_0 = 1, q_0 = 0, p_{-1} = 0, q_{-1} = 1.$ ●

PROOF
The recurrence relations are valid for $n = 3$ and $n = 4$. These cases have been verified above by direct calculation, as were $p_1 = a_1, q_1 = 1,$
$p_2 = a_2a_1 + 1 = a_2p_1 + 1, q_2 = a_2.$

We can now see that the conventions introduced at the end of the statement of the theorem are designed to ensure that the recurrence relations are valid for $n = 1$ and $n = 2$, for algebraic convenience.

If we study the calculations involved in the evaluation of the first four convergents, we can see that they have not used the assumption that the values of the partial quotients a_k are integers. They can be any non-zero numbers, rational or otherwise. We make use of this by considering the kth convergent

$$c_k = a_1 + \frac{1}{a_2+} \frac{1}{a_3+} \cdots \frac{1}{a_k},$$

and replacing a_k by $a_k + \dfrac{1}{a_{k+1}}$ to obtain c_{k+1}.

Now suppose that the recurrence relations are valid for $n = k$. We then have

$$c_k = \frac{p_k}{q_k} = \frac{a_k p_{k-1} + p_{k-2}}{a_k q_{k-1} + q_{k-2}}.$$

As explained above, replacing a_k by $a_k + \dfrac{1}{a_{k+1}}$ gives c_{k+1}. So

$$c_{k+1} = \frac{p_{k+1}}{q_{k+1}} = \frac{\left(a_k + \dfrac{1}{a_{k+1}}\right) p_{k-1} + p_{k-2}}{\left(a_k + \dfrac{1}{a_{k+1}}\right) q_{k-1} + q_{k-2}}.$$

Multiplying the numerator and the denominator by a_{k+1} and rearranging the terms gives

$$c_{k+1} = \frac{p_{k+1}}{q_{k+1}} = \frac{a_{k+1}(a_k p_{k-1} + p_{k-2}) + p_{k-1}}{a_{k+1}(a_k q_{k-1} + q_{k-2}) + q_{k-1}} = \frac{a_{k+1} p_k + p_{k-1}}{a_{k+1} q_k + q_{k-1}},$$

using the inductive hypothesis. This implies that both the recurrence relations are valid for $n = k + 1$, and hence proves the theorem by induction.   ●

These recurrence relations enable us to calculate successive convergents starting from the left of the continued fraction, rather than having to work out each one separately by unscrambling each continued fraction from the right-hand end. We can set out the calculations systematically in the form of a table. We shall use the first illustration in this section.

k	-1	0	1	2	3	4	5
a_k			1	1	4	2	7
p_k	0	1	1	2	9	20	149
q_k	1	0	1	1	5	11	82

Let us illustrate the use of the recurrence relations from this table by considering $k = 3, 4, 5$ in turn.

$$p_3 = a_3 p_2 + p_1 = 4 \times 2 + 1 = 9,$$
$$q_3 = a_3 q_2 + q_1 = 4 \times 1 + 1 = 5,$$
$$p_4 = a_4 p_3 + p_2 = 2 \times 9 + 2 = 20,$$
$$q_4 = a_4 q_3 + q_2 = 2 \times 5 + 1 = 11,$$
$$p_5 = a_5 p_4 + p_3 = 7 \times 20 + 9 = 149,$$
$$q_5 = a_5 q_4 + q_3 = 7 \times 11 + 5 = 82.$$

Consider the last two equations from a visual point of view on the table, and suppose that we had completed the calculations as far as p_4 and q_4. To calculate p_5 we start at a_5, so we move to the 7. Now move left diagonally down one row (20) and multiply, and then across to the left (9) and add. To calculate q_5 we start at a_5, so move to the 7 again. Now move left diagonally down two rows (11) and multiply, and then across to the left (5) and add. One soon becomes accustomed to this systematic approach and it enables the convergents to be calculated very quickly.

It was pointed out above that when we convert a continued fraction into its rational number, we obtain a fraction in its lowest terms, as in this illustration. In fact we notice from this example that for all values of $n \geq 1$ in the table p_k and q_k have no common factors. We shall now prove this result.

● Proposition 4

(i) For all $n \in \mathbb{N}$ we have

$$p_n q_{n-1} - q_n p_{n-1} = (-1)^n.$$

(ii) For all $n \in \mathbb{N}$ the highest common factor of p_n and q_n is equal to 1. ●

PROOF

(i) From the calculation of the first four convergents earlier in this section the result can be verified for small values of n. (The algebra is fairly involved for $n = 3$ and $n = 4$.) For $n = 1$ the conventions introduced in Proposition 3 tell us that

$$p_1 q_0 - q_1 p_0 = a_1 \times 0 - 1 \times 1 = -1 = (-1)^1,$$

so the equation is satisfied for $n = 1$.

Suppose that the equation is valid for $n - 1$. Using the recurrence relations of proposition 3 gives

$$p_n q_{n-1} - q_n p_{n-1} = (a_n p_{n-1} + p_{n-2})q_{n-1} - (a_n q_{n-1} + q_{n-2})p_{n-1}$$
$$= -(p_{n-1}q_{n-2} - q_{n-1}p_{n-2}) = (-1)^n,$$

using the inductive hypothesis. Hence, the equation is valid for all $n \in \mathbb{N}$ by induction.

(ii) Using the equation tells us that if d is a positive divisor of p_n and of q_n then d is a divisor of the left-hand side, and so is a divisor of $(-1)^n$. Hence, $d = 1$ and p_n and q_n therefore have no common factors. ●

Finally, let us look briefly at the use of continued fractions as approximations. If we have a number A, and we take as an approximation a number A_n with the first n decimal places equal to those of A, we have

$$A_n \leq A \leq A_n + \frac{1}{10^n}.$$

Now A_n can be expressed as a fraction p/q with denominator $q = 10^n$, so that $|A - A_n| \leq 1/q$. If we now consider convergents for the continued fraction for A it can be shown that for any n, A always lies between the two successive convergents c_n and c_{n+1}. This tells us that

$$|A - c_n| \leq |c_{n+1} - c_n| = \left| \frac{p_{n+1}}{q_{n+1}} - \frac{p_n}{q_n} \right| = \frac{1}{q_{n+1}q_n},$$

using the result of Proposition 4. So if we now use p/q to denote c_n, we have $|A - c_n| \leq 1/q^2$, because $q_{n+1} > q_n$ from the recurrence relation. So n places along the continued fraction will give a much better approximation than n decimal places. Let us illustrate this with reference to the continued fraction for π. This is an infinite continued fraction for which the first few places were worked out by Lambert in 1770. He showed that

$$\pi = [3; 7, 15, 1, 292, 1, 1, 1, 2, 1, 3, 1, 14, 2, 1, 1, 2, 2, 2, 2, 1, 84, 2, \ldots].$$

There is no discernible regular pattern to the partial quotients. If we evaluate the first few convergents using the tabular method we obtain

k	-1	0	1	2	3	4
a_k			3	7	15	1
p_k	0	1	3	22	333	355
q_k	1	0	1	7	106	113

The second convergent is $\frac{22}{7}$, and the fourth is $\frac{355}{113}$. You will have used 22/7 as an approximation for π for a long time no doubt, and its first two places agree with those of π. For the fourth convergent the first six decimal places agree, in fact the error is at most 3 in the seventh decimal place. The fourth has been chosen here because it is easy to remember, as it contains two 1s, two 3s and two 5s.

EXERCISES 3.3

1. Evaluate the continued fractions for the following rational numbers.

$$\frac{89}{55}, \quad \frac{1101011}{1001010}, \quad \frac{1393}{972}, \quad \frac{6961}{972}, \quad \frac{169}{70}.$$

2. Practice the tabular method by choosing a few sets of six integers, use them as partial quotients for continued fractions, and calculate the associated rational number, and the intermediate convergents.

3. Euler, whose book is referred to above, found several continued fractions connected with the exponential number e. In particular

$$e - 1 = [1; 1, 2, 1, 1, 4, 1, 1, 6, 1, 1, 8, \ldots],$$
$$\frac{e-1}{2} = [0; 1, 6, 10, 14, 18, 22, \ldots],$$

where the second one continues with the partial quotients increasing by 4 each time. Calculate the first seven convergents of each, using the tabular method, and show that the second continued fraction gives better approximations to the number e. [This relates to the fact that the partial quotients are larger than those in the first expansion.]

4. Write a computer program to calculate the continued fraction and the convergents for a rational number. You can base it on the program segment for the Euclidean Algorithm given in §2.4. The recurrence relations in Proposition 3 can be used to generate the numerators and denominators for the successive convergents. The initial values given in the statement of Proposition 3 will be useful for starting data.

Summary

One of the main purposes of this chapter has been to illustrate two of the methods used in many places in mathematics.

The first is the axiomatic method, which seeks to formalize parts of mathematics which have perhaps been explored less formally or on an empirical basis. It has been part of the methodology of mathematics since the time of Euclid, and it experienced a major resurgence during the 18th and 19th centuries with the invention of different kinds of geometry and algebra.

The second approach to rational numbers again exhibits a general methodology, that of creating new mathematical structures, especially using the mechanism of equivalence relations. This is used in many branches of analysis, algebra and geometry, and particularly in modern topology. As one example, we can cite the investigation of the symmetries of regular solid shapes.

In applying these general approaches to a topic as familiar as that of fractions, it is hoped that some feeling for the methodology will be acquired, so that later applications in less familiar areas will be facilitated, in particular when group theory is studied.

The final section of this chapter, on continued fractions, has barely scratched the surface of this fascinating subject. It is intended to give a new perspective on the topic of fractions, which most readers will have studied for many years. It contains some more proofs by induction, and provides some new types of calculation which students can enjoy. In particular, it contains examples of algorithms, another general method of increasing importance in present day mathematical research.

EXERCISES ON CHAPTER 3

1. Consider the simultaneous linear equations $a_1 x + b_1 y = c_1$ and $a_2 x + b_2 y = c_2$. Analyse the procedure for solving these equations for x and y in terms of the axioms for $\mathbb{Q}$, in a similar way to Example 1.

2. Prove that for all $a \in \mathbb{Q}$ the additive inverse identified in axiom A4 is unique. Prove that for $a \neq 0$ the multiplicative inverse (axiom M4) is unique.

3. Show that the rational number system as constructed in §3.2, with addition as defined before Proposition 2, and with multiplication defined similarly (Exercise 1 showing that multiplication is 'well-defined'), satisfies the axioms for a field as given in §3.1. You will need to use the fact that all the axioms except M4 are true with $\mathbb{Z}$ replacing $\mathbb{Q}$. Note that the definitions of the operations themselves ensure that A1 and M1 are satisfied.

   (The verification of axioms—particularly associativity—is regarded as tedious but necessary by many mathematicians. It is suggested that at the least you should describe the equivalence classes of fractions which act as additive and multiplicative identities and inverses, and verify axioms A3, A4, M3 and M4.)

4. Explain how the result of Proposition 4(i) enables us to write down a solution of the equation $p_n x + q_n y = 1$, where x and y are integers. Find all the integer solutions of the equation $1393x + 972y = 1$. (Refer to Example 8 in Chapter 2 for a method of finding all solutions once one has been determined.)

$4 \bullet$ Inequalities

So far we have concentrated on the arithmetic operations of numbers and the associated algebraic properties. The other important properties we have to consider are those relating to the ordering of numbers; the idea that one is larger than another. This leads to the problem of solving an algebraic inequality such as $x^2 - 3x + 4 < 0$. From school mathematics solving equations is a familiar activity, but manipulating inequalities less so. We shall therefore start with the basic rules for manipulating inequalities, and then look at several methods for solving them. The solutions of the inequalities we consider will be sets of real numbers, which are not formally defined until Chapter 5. It is sufficient here to rely on the intuition provided by reference to the standard number line, where larger numbers are situated to the right of smaller ones.

4.1 The Basic Rules for Inequalities

In Chapter 3 we listed the rules for algebra as a system of axioms. The rules for inequalities concern interaction with the operations of arithmetic. We formulate them as axioms for ordering as follows, where in each case the variables a, b, c can take all values in the real numbers. It is axioms O3 and O4 which are used in the algebraic manipulation of inequalities.

O1 (Trichotomy) If $a \neq b$ then either $a < b$ or $b < a$.

O2 (Transitivity) If $a < b$ and $b < c$ then $a < c$.

O3 (Translation) If $a < b$ then $a + c < b + c$.

O4 If $a < b$ and $c > 0$ then $ac < bc$.

It is very important to note that O4 requires the multiplier c to be positive. An example will illustrate this. If we take $2 < 5$ and multiply both sides by 3 we obtain $6 < 15$, which is true. If on the other hand we try to multiply by -3 we would obtain $-6 < -15$ which is false. In fact the inequality is reversed, so that $-6 > -15$. This always happens, as the following result demonstrates.

$\bullet$ Proposition I

If $a < b$ and $c < 0$ then $ac > bc$. $\bullet$

PROOF
Since $c < 0$ we can add $-c$ to both sides and use O3 to obtain $0 < -c$. We can now use O4 to deduce that $a \times (-c) < b \times (-c)$, so that $-ac < -bc$. Finally, we can add ac and bc to both sides using O3 to obtain $bc < ac$, which is equivalent to $ac > bc$. $\bullet$

We can illustrate O3, O4 and this last result on a number line, where we interpret addition as a shift and multiplication as a magnification. The first two figures illustrate the idea of adding to both sides of an inequality, where the number added can be positive or negative.

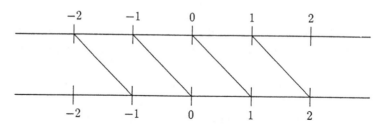

Fig 4.1 If $a < b$ then $a + 1 < b + 1$.

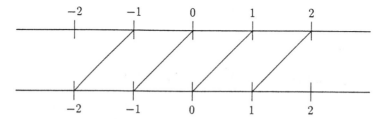

Fig 4.2 If $a < b$ then $a - 1 < b - 1$.

The second pair of figures illustrates multiplying. In the first case, order is preserved, while in the second case order is reversed, corresponding to the result of Proposition 1.

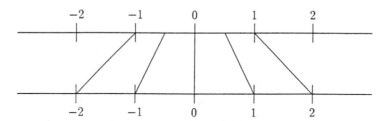

Fig 4.3 If $a < b$ then $2a < 2b$.

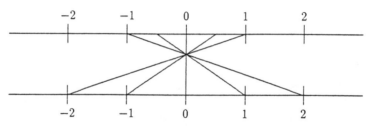

Fig 4.4 If $a < b$ then $(-2)a > (-2)b$.

Example I

Show that if $a < 0$ then $a^2 > 0$.

We use the result of Proposition 1 in the particular case when $b = a$ and $c = a$. This then gives $a \times a > 0 \times 0 = 0$, using Example 5 of §2.3.

Many inequalities involve the magnitude of a number, representing its size without regard to sign. We use a convenient notation for this, described as follows.

● Definition I

For any number x we define the modulus (or magnitude) of x by

$$|x| = \begin{cases} x, & \text{if } x \geq 0; \\ -x, & \text{if } x < 0. \end{cases}$$

TUTORIAL PROBLEM I

Draw the graphs of $y = |x|$, $y = x + |x|$, $y = |2x - 3|$ and explore other graphs involving the modulus.

Example 2

Show that for all a, b, $|a + b| \leq |a| + |b|$. This result is known as the triangle inequality, for reasons that will become clear in Chapter 6.

A useful property of the modulus is that $|a|$ is the positive square root of a^2, and we use that here, as follows,

$$(|a + b|)^2 = (a + b)^2 = a^2 + b^2 + 2ab = |a|^2 + |b|^2 + 2ab \leq |a|^2 + |b|^2 + 2|a||b|$$
$$= (|a| + |b|)^2.$$

Taking the positive square root of both ends of this chain of relations gives the result. Now we shall have equality if and only if $ab = |a||b|$, i.e. if and only if $ab \geq 0$. So either a and/or b is zero or else a and b must both be of the same sign.

TUTORIAL PROBLEM 2

Construct a proof by induction that
$|a_1 + a_2 + \ldots + a_n| \leq |a_1| + |a_2| + \ldots |a_n|$. The result is trivial for $n = 1$, and Example 2 gives the case $n = 2$. The inductive step should use the case $n = 2$ to deduce that $|a_1 + a_2 + \ldots + a_n| \leq |a_1 + a_2 + \ldots + a_{n-1}| + |a_n|$ and then an appropriate inductive hypothesis should be invoked.

EXERCISES 4.1

1. If $0 < a < b$, use axiom O4 to deduce that $a^2 < b^2$. [Hint: take $c = a$ and $c = b$ in turn.]

2. Find examples, involving negative numbers, where (i) $a < b$ and $a^2 < b^2$, (ii) $a < b$ and $a^2 > b^2$. Try to discover rules to determine which of these cases occurs in general.

3. Suppose $a < b$ and $c < d$. Use axiom O3 to show that $a + c < b + d$.

4. Suppose $a < b$ and $c < d$. Find some numerical examples to show that either $a - c < b - d$ or $a - c > b - d$ can occur, depending on the numbers involved.

5. Show that if $0 < a < b$ then $1/b < 1/a$ (use axiom O4). Investigate this situation in the case where $a < 0$ or $b < 0$, or both.

4.2 Solving Inequalities Graphically

If we have an inequality of the form $f(x) < g(x)$ to solve, we can draw the graphs of $f(x)$ and $g(x)$ and investigate where the first lies below the second. Generally, this will be achieved by finding the points of intersection, which involves solving the equation $f(x) = g(x)$. Solving equations is a more familiar procedure. We shall illustrate the method with some examples.

Example 3

Solve the inequality $x^2 - 4x + 1 < 3$.

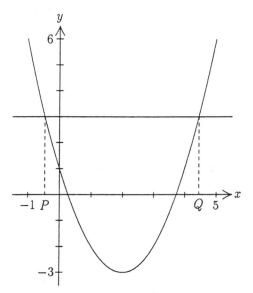

Fig 4.5 Solving the inequality $x^2 - 4x + 1 < 3$ graphically.

Figure 4.5 shows the graphs of $y = x^2 - 4x + 1$ and $y = 3$ on the same diagram, from which it can be seen that the solution of the inequality consists of all the values

of x between the points P and Q. These can be found by solving the equation
$x^2 - 4x + 1 = 3$. This is a quadratic equation whose solutions are $x = 2 \pm \sqrt{6}$, and
so the solution of the inequality is the set of all values of x satisfying
$2 - \sqrt{6} < x < 2 + \sqrt{6}$.

Example 4

Solve the inequality $\dfrac{2 - x}{3 + x} < 4$.

Using the same method as Example 3 we draw the graphs of $y = (2 - x)/(3 + x)$
and $y = 4$ on the same axes. It can then be seen, in Fig 4.6, that the solution consists
of all the values of x to the left of the vertical asymptote, together with all the values
to the right of the point P. The vertical asymptote is at $x = -3$, and the point P is
found by solving the equation $\frac{2-x}{3+x} = 4$, which gives $x = -2$. So the solution of the
inequality is the set of all values of x satisfying $x < -3$ or $x > -2$.

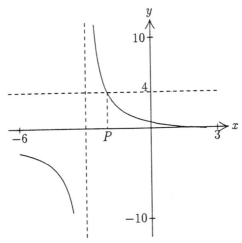

Fig 4.6 Solving the inequality $\frac{2-x}{3+x} < 4$ graphically.

Example 5

Solve the inequality $\sin x > x^2 - 2$.

Once again we draw two graphs on the same axes, namely $y = \sin x$ and $y = x^2 - 2$.

We can see from the part of the graphs that we have drawn that there are two
points, P and Q, where they intersect. We now observe that because the quadratic
graph increases to the right and also to the left, and because $-1 \le \sin x \le 1$ for all
x, there can be no other points of intersection. This means that the solution consists
of all those values of x lying between P and Q. In this case we cannot solve the
equation $\sin x = x^2 - 2$ exactly, as we were able to do with the equations in
Examples 2 and 3 in this section. We must use a graphical or numerical method, and
in this case the computer package Graphical Calculus was used, with its Zoom
facility enabling us to obtain a good approximation to the points of intersection. In

this way we can say that the solution of the inequality is approximately the set of values of x satisfying $-1.0614 < x < 1.7285$, where the left-hand number has been rounded down and the right-hand one rounded up. This gives a slightly larger interval than the true one, which is what is needed sometimes in problems of estimation.

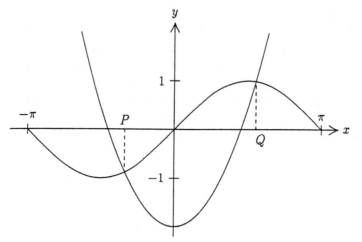

Fig 4.7 Solving the inequality $\sin x > x^2 - 2$ graphically.

EXERCISES 4.2

1. Solve the following inequalities graphically,
   (i) $x^2 - 4x + 2 > 3$, (ii) $x^2 - 4x + 2 \leq -2$, (iii) $x^2 - 4x + 2 < 4$,
   (iv) $x^2 - 4x + 2 < -3$.

2. Solve the following inequality graphically,

$$-2 < \frac{x+2}{2x-1} < 3.$$

3. Solve the inequality $\cos x < 3 - x$. Give your answers correct to four decimal places if you have a suitable calculator or computer available.

4.3 Solving Inequalities Algebraically

The basic rules for manipulating inequalities were discussed in §4.1. They show some similarities to the procedures for solving equations and so we shall see some familiar processes taking place in the following examples. We shall first deal with the inequalities in Examples 2 and 3 above from an algebraic point of view.

Example 6

Solve the inequality $x^2 - 4x + 1 < 3$.

Completing the square gives $(x-2)^2 - 3 < 3$. Adding 3 to both sides then gives $(x-2)^2 < 6$. We now have an inequality of the form $t^2 < A$, where A is a positive real number. Unlike an equation, we cannot simply take the square root of both sides. If $t^2 < 4$ it is incorrect to deduce that $t < 2$, because for example $t = -5$ satisfies the second inequality but not the first. The correct deduction is $-2 < t$ and $t < 2$, which we conventionally write together as $-2 < t < 2$. Returning to the example, we can now say that $(x-2)^2 < 6$ implies that $-\sqrt{6} < (x-2) < \sqrt{6}$. Adding 2 throughout finally gives $2 - \sqrt{6} < x < 2 + \sqrt{6}$ as the solution.

Note that we could write $|t| < 2$ in place of $-2 < t < 2$, and also $|x-2| < \sqrt{6}$ as equivalent to $-\sqrt{6} < (x-2) < \sqrt{6}$.

TUTORIAL PROBLEM 3

Use the method of completing the square shown in Example 6 to prove that the general quadratic expression $ax^2 + bx + c$ is positive for all real x if and only if $b^2 < 4ac$ and $a > 0$. Find analogous conditions for the quadratic to be negative for all real x. Find the solutions of $ax^2 + bx + c \geq 0$ in the case when $b^2 \geq 4ac$. (Remember to consider both $a > 0$ and $a < 0$.)

Example 7

Solve the inequality $\dfrac{2-x}{3+x} < 4$.

It is tempting to perform the same algebraic operation we would if we had an equation, namely to multiply both sides by $(3 + x)$. Let us do that and see what happens. We then obtain $2 - x < 4(3 + x)$, and so $2 - x < 12 + 4x$. Axiom O3 tells us that we can add to and subtract from both sides of an inequality, and so we obtain $2 - 12 < 4x + x$, i.e. $-10 < 5x$, giving $x > -2$. However, if we look back to Example 4 where we solved this inequality graphically we observe that we have obtained only part of the solution. We appear to have lost the other part, $x < -3$. What has happened is that we have failed to use the important part of axiom O4 which says that an inequality is preserved if the multiplier is positive. Proposition 1 then established that the inequality is reversed if the multiplier is negative. We have not considered here the sign of $(3 + x)$, the multiplier. We shall now do that to obtain the complete solution algebraically.

Case 1: $(3 + x) > 0$. In this case the algebra above is correct, giving $x > -2$.

Case 2: $(3 + x) < 0$. Multiplying by the negative quantity $(3 + x)$ now gives $2 - x > 4(3 + x)$, i.e. the inequality has been reversed. This rearranges to give $x < -2$. But we have the initial condition, equivalent to $x < -3$. So the solution to Case 2 is $x < -2$ and $x < -3$. The set of values of x satisfying both these inequalities simultaneously gives $x < -3$.

To give a complete explanation of Case 1 we really should have taken the initial condition there into account, so we conclude that $x > -2$ and $x > -3$, which this time reduces to $x > -2$.

Example 8

Solve the inequality $\dfrac{3 - 2x}{x^2 + 1} < 3 - x$.

In dealing with equations we are accustomed to the algebraic strategy of collecting all the terms on one side. This is a strategy which is also useful here, and so we obtain

$$\frac{3 - 2x}{x^2 + 1} - (3 - x) < 0.$$

It is often a good idea to minimize the possibility of errors arising through having several minus signs, and with this in mind we rewrite the last inequality as

$$\frac{3 - 2x}{x^2 + 1} + (x - 3) < 0.$$

Putting this over a common denominator gives

$$\frac{3 - 2x + (x - 3)(x^2 + 1)}{x^2 + 1} < 0.$$

We can now simplify the algebra by realizing that $x^2 + 1$ is positive for all values of x, and so the expression will be negative if and only if the numerator is negative. So we now have to solve $3 - 2x + (x - 3)(x^2 + 1) < 0$. (Note that we could also have obtained this by multiplying both sides of the original inequality by $x^2 + 1$, which is positive for all x.)

Multiplying out the brackets and collecting terms gives $x(x^2 - 3x - 1) < 0$. We now have an expression which is a product of two factors, and so will be negative if and only if the two factors have opposite sign. We will therefore consider two cases.

Case 1: $x > 0$ and $x^2 - 3x - 1 < 0$.

Completing the square for the quadratic gives $\left(x - \frac{3}{2}\right)^2 - \frac{13}{4}$, and so this case gives

$$x > 0 \quad \text{and} \quad \left(x - \frac{3}{2}\right)^2 - \frac{13}{4} < 0,$$

$$x > 0 \quad \text{and} \quad \left(x - \frac{3}{2}\right)^2 < \frac{13}{4},$$

$$x > 0 \quad \text{and} \quad \frac{3}{2} - \frac{\sqrt{13}}{2} < x < \frac{3}{2} + \frac{\sqrt{13}}{2}.$$

Now $\dfrac{3}{2} - \dfrac{\sqrt{13}}{2}$ is a negative number, so that the two conditions of this case amalgamate to give

$$0 < x < \frac{3}{2} + \frac{\sqrt{13}}{2}.$$

Case 2: $x < 0$ and $x^2 - 3x - 1 > 0$.

This time completing the square and rearranging gives

$$x < 0 \quad \text{and} \quad \left(x - \frac{3}{2}\right)^2 > \frac{13}{4}.$$

The second inequality is of the form $t^2 > A$ and, as in Example 6, a numerical illustration will help. If we have $t^2 > 4$ then t could be positive or negative, giving two possibilities, namely $t > 2$ or $t < -2$. So Case 2 becomes

$$x < 0 \quad \text{and} \quad \left(x < \frac{3}{2} - \frac{\sqrt{13}}{2} \quad \text{or} \quad x > \frac{3}{2} + \frac{\sqrt{13}}{2}\right).$$

The fact that x is negative means that the third inequality cannot be satisfied, and so this case reduces to

$$x < \frac{3}{2} - \frac{\sqrt{13}}{2}.$$

Finally, putting together Cases 1 and 2 gives the complete solution as

$$x < \frac{3}{2} - \frac{\sqrt{13}}{2} \quad \text{or} \quad 0 < x < \frac{3}{2} + \frac{\sqrt{13}}{2}.$$

Both the algebra and the logic in this example offer some difficulties, and it is advisable to check with a graphical method if possible, using a graphical calculator or a computer package if available.

In the final example in this section we use a combination of the graphical and algebraic approaches.

Example 9

Solve the inequality $2|x| > |2x - 5| + |x + 1|$.

We first draw graphs of each side of the inequality on the same axes, either 'by hand' or using a graphical calculator or a computer (Fig 4.8).

The dashed graph represents $y = 2|x|$ and the solid graph represents $y = |2x - 5| + |x + 1|$. The latter graph changes direction where $x = -1$ and $x = 5/2$, i.e. when $|x + 1| = 0$ and where $|2x - 5| = 0$. From the graph we can see that the solution of the inequality consists of the numbers between the points P and Q. In the neighbourhood of the point P, $|2x - 5| + |x + 1| = -(2x - 5) + (x + 1) = -x + 6$. So to find P we have to solve the equation $-x + 6 = 2x$, giving $x = 2$. In the neighbourhood of Q, $|2x - 5| + |x + 1| = (2x - 5) + (x + 1) = 3x - 4$, so we have to solve $3x - 4 = 2x$, giving $x = 4$. The solution is therefore the set of values of x satisfying $2 < x < 4$.

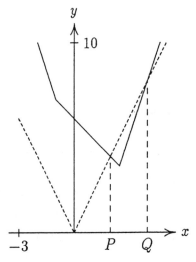

Fig 4.8 Solving the inequality $2|x| > |2x-5| + |x+1|$.

1. Solve the inequalities in the first two exercises of §4.2 algebraically.
2. Solve the following inequality algebraically and confirm your results graphically.

$$x - 4 > \frac{7x - 8}{x^2 + 2}.$$

4.4 A Tabular Approach to Inequalities

We saw that Example 7 involved a consideration of the signs of factors in a product. In this section we shall look at some more examples of this kind, adopting a systematic analysis of sign changes. We shall consider examples in which we assume that some inequality has been rearranged so that is has the form $f(x) > 0$, where the formula for $f(x)$ involves the product and quotient of a number of factors. In the most straightforward examples these factors are linear, and we shall begin with such a situation.

Example 10

Solve the inequality $\dfrac{(x+3)(x-2)}{(x+1)(2x-1)} > 0$.

There are four factors in the rational function here, and the changes of sign occur at $x = -3, -1, 1/2$ and 2. The function will be positive if and only if an even number of these factors is negative. We set out a table in which we mark divisions relating to the sign changes, and an indication between these divisions of the sign of each factor.

		−3		−1		1/2		2	
$x + 3$		−		+		+		+	+
$x + 1$		−		−		+		+	+
$2x - 1$		−		−		−		+	+
$x - 2$		−		−		−		−	+
TOTAL		+		−		+		−	+

Fig 4.9 Tabular solution of the inequality $\frac{(x+3)(x-2)}{(x+1)(2x-1)} > 0$.

We conclude from this table that the inequality is satisfied if $x < -3$ or $-1 < x < 1/2$ or $x > 2$.

In fact, any inequality of the form $R(x) > 0$, where $R(x)$ is a rational function, can in theory be solved by this tabular method. Any polynomial factorizes into a product of factors which are either linear or non-factorizable quadratic. The factorization can be arranged so that these quadratic factors are positive for all x, and so can be omitted, as we did with $x^2 + 1$ in Example 7. The problem therefore reduces to one of a product and quotient of linear factors which can therefore be dealt with exactly like Example 10.

For the other example in this section we shall consider an inequality involving a non-rational function.

Example 11

Solve the inequality $(x - 1)(x + 2) \sin x < 0$.

Again we consider the values of x where the expressions change sign. These are $x = 1$, $x = -2$ and integer multiples of π. We construct a table of the kind used in Example 10.

		-3π		-2π		$-\pi$	-2		0		1		π		2π		3π
$\sin x$		−		+		−		−		+		+		−		+
$x - 1$		−		−		−		−		−		+		+		+
$x + 2$		−		−		−		+		+		+		+		+
TOTAL		−		+		−		+		−		+		−		+

Fig 4.10 Tabular solution of the inequality $(x-1)(x+2) \sin x < 0$.

Outside the confines of this table the product of the three factors will clearly alternate in sign, changing every time an integer multiple of π is encountered. So we can say that the values of x which make the product negative satisfy $-\pi < x < -2$ or $0 < x < 1$ or $n\pi < x < (n+1)\pi$, where n can be any odd integer satisfying $n \geq 1$ or $n \leq -3$.

EXERCISES 4.4

1. Solve the following inequalities using the tabular method,

    (i) $\dfrac{x+2}{(x-1)(x-3)} < 0$, (ii) $(x-2)(x+3)(x^2-5x+6) < 0$,

    (iii) $\dfrac{x^2-4}{x^2-5x+6} > 0$, (iv) $(x+3) \cos x > 0$, (v) $\sin x \ln x < 0$.

4.5 Increasing and Decreasing Functions

A function $f(x)$ is an increasing function if the value of $f(x)$ increases as the variable x increases. This means that the graph moves up as you move along the x–axis, with the normal orientation of Cartesian axes. These ideas can be stated in terms of inequality as follows.

A function $f(x)$ is said to be increasing for x between a and b if for all values of x and y satisfying $a \le x < y \le b$ we have $f(x) \le f(y)$. Notice the use of the $\le$ symbol in the last inequality. This allows the possibility of equality in some cases, which means that the graph of $f(x)$ could have horizontal portions, as with a distance–time graph when the moving object has periods of rest.

Common examples of increasing functions are the exponential and logarithmic functions, the cube function and the tangent function between $-\pi/2$ and $\pi/2$. Using increasing functions enables us to deduce complicated inequalities from simpler ones. For example, since $x^2 > x$ for all $x > 1$ it follows that $\exp(x^2) > \exp(x)$ for all $x > 1$.

Example 12

Solve the inequality $\exp((x-1)(x+2) \sin x) < 1$.

Since the logarithmic function is increasing, the inequality will be satisfied if and only if $\ln(\exp((x-1)(x+2) \sin x)) < \ln 1$. Using the fact that logarithm and exponential are inverse functions, and that $\ln 1 = 0$, this reduces to the inequality $(x-1)(x+2) \sin x < 0$, which was solved in Example 9.

Example 13

Solve the inequality $\ln(\frac{1}{2}x^2 + x + 1) > 0$.

The exponential function is increasing, so this inequality is equivalent to $\frac{1}{2}x^2 + x + 1 > e^0 = 1$. Using the method of completing the square discussed in Examples 6 and 8 gives the solution as $x > 0$ or $x < -2$. The detailed calculations are left to the reader.

A function $f(x)$ is said to be decreasing for x between a and b if, for all values of x and y satisfying $a \le x < y \le b$, we have $f(x) \ge f(y)$. Pictorially the graph goes down as you move along the x–axis in the positive direction. Common examples are e^{-x}, $\cos x$ for x between 0 and π, $1/x$ for $x > 0$. The reciprocal function is often encountered in dealing with inequalities. So, for example, because $x^3 > x^2$ for $x > 1$ it follows that $1/x^3 < 1/x^2$ for $x > 1$, since the reciprocal function is decreasing.

EXERCISES 4.5

1. Use the fact that the logarithmic function is increasing to solve the inequality $e^x < 2^{x^2}$.

2. Solve the inequality $((x-1)(x+2))^3 < ((x-3)(x+5))^3$.

3. Solve the inequality $\ln\left(\dfrac{2-x}{12+4x}\right) < 0$, using the results of Example 7.

Summary

The purpose of this chapter has been to explore a variety of methods of finding the solutions of inequalities. When the tabular method is applicable it gives a straightforward approach, but it depends on the function being given as a product or quotient of expressions, for each of which the points of sign change are known or can easily be found. The graphical method is in some ways the most flexible, as it gives a picture of the situation as well as reducing the problem to the solution of equations. Drawing graphs 'by hand' is sometimes tricky, but with a graphical calculator or a graph plotting computer package the pictorial aspect is automatic. There is still some mathematical thinking to be done however, for one must be certain that all the points of intersection of the graphs used are included in the part of the graph shown on the screen. If this is not so, it is possible in most cases to change the domain on the calculator or computer, but if one forgets to investigate this matter then it is easy to miss parts of the solution. The algebraic approach in §4.3 is in some ways the most demanding, especially where it splits into several cases according to the sign of any multiplying factor used. Ideally, one will use an approach where graphical and algebraic skills can be employed in conjunction and reinforce one another.

EXERCISES ON CHAPTER 4

1. Using only Axiom O3 prove that if $a < b$ then $-b < -a$.

2. Prove that if $a < b < 0$ then $b^2 < a^2$.

3. If $0 < a < b$, prove by mathematical induction that $a^n < b^n$ for all positive integers n.

4. Solve the following inequality graphically
$$-1 < \frac{2x+3}{x-4} < 5.$$

Rearrange each side of the inequality in a form suitable for solution by the tabular method, and implement the method to check the results of the graphical approach.

5. Solve the inequality $-\frac{1}{2} < \sin x < \frac{1}{2}$.

5 • The Real Numbers

Having explored some aspects of the integers and the rational numbers in Chapters 2 and 3 we now have to consider the next stage and analyse what it is that distinguishes the real numbers from the rational numbers. One of the methods we have used in previous chapters is to consider the solution of equations, and in discussing the solution of $x^2 = 2$ we highlight a defect of the rational number system, showing that there are gaps. These are not in the form of 'gaping holes', for between any two rational numbers there is always another, so there are no gaps in the form of intervals.

Extending the integers to the rationals was an algebraic matter. Extending the rationals to the reals in quite different, and a conceptually more advanced process. The method we shall use involves a close analysis of statements involving quantifiers, which were introduced in Chapter 1. It is one of several possible approaches to the analysis of the relationships between the real and rational numbers.

5.1 Gaps in the Rational Number System

The Greek problem of incommensurability was referred to in Chapter 3. We can transform this problem into one involving numbers as follows.

If $ABCD$ denotes a square, then Pythagoras' Theorem tells us that the diagonal AC satisfies the equation $AC^2 = 2AB^2$. If the side AB and the diagonal AC were commensurable, it would be possible to find a common unit of measurement. AC would then be, say, m units long, and AB would be, say, n units long. We would therefore have $m^2 = 2n^2$. This is now a statement about numbers rather than lengths. If we divide both sides by n^2 we obtain $(m/n)^2 = 2$, so the problem can be seen as one of investigating whether there is a rational number whose square is equal to 2. We shall prove that this is impossible, using the method of *reductio ad absurdum* described in Chapter 1.

• Proposition I

There is no rational number whose square is equal to 2.

FIRST PROOF
Suppose that there is a rational number m/n whose square is equal to 2. By cancelling, if necessary, we can assume that m/n is a fraction in its lowest terms, i.e. that the integers m and n have no common factors. Now $m^2 = 2n^2$ implies that m^2 is even. Therefore, m is even (Example 1 of Chapter 1) and so we can write $m = 2k$. Hence $(2k)^2 = 2n^2$, i.e. $4k^2 = 2n^2$. This tells us that $n^2 = 2k^2$, so that n^2 is even, implying that n is even. But m and n both being even implies that they do have a

common factor of 2. This is the contradiction which the method of *reductio ad absurdum* entails. The assumption of a rational number whose square is 2 must therefore be false.

SECOND PROOF
Here we use the fact that a prime factor of a square number occurs an even number of times in the factorization. For example $7056 = 84^2 = 2^4.3^2.7^2$. We consider the equation $p^2 = 2q^2$. On the left-hand side the prime factor 2 must occur an even number of times (remember that 0 is even), whereas on the right-hand side 2 will occur an odd number of times. This is a contradiction to the fact that the prime factorization of a number is unique. So it is impossible to have integers p and q which satisfy $p^2 = 2q^2$. ●

TUTORIAL PROBLEM I

> Discuss where such proofs break down if we attempt to show that there is no rational number whose square is equal to 9.

The relationship between measurement and number raises another problem. We are accustomed to representing the real number system by points on a line, most commonly in coordinate geometry. However, the phenomenon of incommensurability is bound to raise serious questions. If the diagonal of a square cannot be measured in terms of the units employed to measure the sides, how can we justify the assumption that, on a line, a point whose distance from a chosen origin is equal to the diagonal of a square will in fact correspond to a number? So are there points on the line which do not correspond to numbers? We might also imagine that other problems could give rise to the reverse question as to whether there could be numbers which would not be associated with points on the line. These are difficult and subtle questions, which occupied philosophers and mathematicians in Greek times and have done so ever since. We shall not pursue them here, but simply note that they point to the desirability of formulating a description of the real number system in purely arithmetic terms, without reference to geometry (other than as analogy).

EXERCISES 5.I

1. Prove that there is no rational number whose square is equal to 3, using the first form of the above proof.
2. Prove that there is no rational number whose square is equal to 12.
3. Prove that there is no rational number r satisfying the equation $r^3 = 4$.

5.2 An Historical Interlude

In this section we shall consider briefly some of the historical background to the development of mathematical descriptions of the real number system.

When Isaac Newton (1642–1727) invented the differential calculus it was found to be a very powerful method for solving problems in many situations involving motion, from ballistics to planetary orbits, and so the calculus gained many strong supporters and exponents. However, the logical foundations came in for considerable criticism in some quarters, with justification, both in terms of the state of mathematics at the time and also in hindsight. Two ideas in particular were given much thought in succeeding years; the nature of a function, and the concept of a limit.

Leonhard Euler (1707–1783) realized that some of the difficulties arose through attempting to interpret the foundations of the calculus geometrically. He viewed the notion of function rather formally in terms of its algebraic representation rather than describing a relationship between numbers. His view of the ratio 0/0, which occurs when we try to calculate the differential coefficient, remained unsatisfactory.

An early realization that a proper theory of limits was needed came through the work of Jean le Rond d'Alembert (1717–1783), but he still considered that the ratio 0/0 could be made meaningful.

Joseph Louis Lagrange (1736–1813) tried to base the calculus on a purely algebraic approach. His view of a function was formal, like that of Euler, but he went further and asserted that an arbitrary function can be expanded as a power series

$$f(a + x) = f(a) + a_1 x + a_2 x^2 + a_3 x^3 + \ldots,$$

that the coefficients could be determined purely algebraically and, in fact, related to the successive differential coefficients of $f(x)$. In fact, the series expansion had been developed by Brook Taylor (1685–1731) in around 1717, but Lagrange sought to admit as functions only those which satisfied his algebraic criteria. Other relationships arising, for example, through graphs having discontinuities or sharp corners would not be admitted as functions in Lagrange's programme. Euler had also done a great deal of work on series, and his profound intuitive insights had led him to many correct conclusions by means of some dubious reasoning. Carl Friedrich Gauss (1777–1855) undertook some rather more careful work in this area, and began to investigate conditions for convergence of some important series.

Significant progress was made by Augustin-Louis Cauchy (1789–1857), who also realized that a proper theory of limits was required, and began to develop such a theory, formulating definitions with considerably more precision than had been achieved previously. Much of this work appeared in a famous book *Cours d'Analyse* in 1821.

One strand of development which came to play an increasingly important role related to a problem from physics—that of determining the motion of a vibrating string with an arbitrary starting shape. Euler and d'Alembert had both investigated this problem, and in fact one of the features was that the initial configuration need not be representable as a function as understood by Lagrange. For example, a string plucked at its middle point could have an initial shape with a sharp corner at the middle. Brook Taylor, and later Euler, investigated solutions of the problem involving sums of trigonometric functions rather than powers of x as in Lagrange's theory. Two other important people in this work were the Bernoullis, John (1667–1748) and Daniel (1700–1782).

A related problem was that of heat conduction, and Joseph Fourier (1768–1830) investigated this, culminating in 1822 with the publication of his book *Théorie Analytique de la Chaleur*. He too was concerned with expressions involving sums of trigonometric functions. These have become known as Fourier Series, and they represent periodic functions in the form

$$f(x) = a_0 + a_1 \cos x + b_1 \sin x + a_2 \cos 2x + b_2 \sin 2x + \dots.$$

They have been widely used in investigating waveforms of all kinds, especially sound waves, and they form a theoretical basis for electronic sound production that can imitate conventional musical instruments and produce other sounds as well—the modern synthesizer.

It was recognized quite soon that the Lagrange notion of a function was inadequate, as Fourier Series could represent graphs with many discontinuities, composed of disconnected, unrelated pieces. Mathematicians such as Peter Lejeune Dirichlet (1805–1859) and Bernhard Riemann (1826–1866) undertook a considerable amount of research into the conditions under which a given function can be represented by a Fourier Series. Because of the discontinuities which can occur at individual points, or values of x, it began to emerge that investigation was needed into the structure of numbers themselves.

Lagrange's notion of a function had suggested that a function with a continuous graph should be differentiable. Subsequent developments suggested isolated exceptions, at corner points for example, but this idea was completely overturned when examples were constructed of functions which were continuous everywhere but differentiable nowhere. This was done geometrically in about 1830 by Bernard Bolzano (1781–1848), and then analytically in 1861 by Karl Weierstrass (1815–1897). The latter example again involved a sum of trigonometric functions. Weierstrass realized how necessary it was to base theories of limits upon a precise analysis of the number system, and the nature of irrational numbers in particular, and this he began to investigate. Many mathematicians took part in this development during the latter part of the 19th century, among them Eduard Heine (1821–1881) and in particular Georg Cantor (1845–1918) and Richard Dedekind (1831–1916). It is these latter two mathematicians whose names we chiefly associate with the development of the theory of real numbers, and the analysis of the relationships between rational and irrational numbers. It is interesting to note that Cantor himself worked on problems involving trigonometric series, which serves to emphasize their importance in this development. He also invented a system of infinite (transfinite) numbers, and began to undertake research into many aspects of the structure of the number line, leading to the growth of topology in the first part of the 20th century.

5.3 Bounded Sets

We shall consider the step from rational numbers to the real number system through the notion of a bounded set. To visualize the ideas being introduced we shall rely on the number line, in spite of the logical reservations expressed in §5.1. In fact, we shall describe the concepts involved verbally, symbolically and geometrically. For the purposes of illustration the sets shown in the figures below appear as intervals. This will not be the case in general, and we shall give examples not involving intervals.

● *Definition 1*

A set S of numbers is said to be bounded above if there is a number which is greater than all members of S. Symbolically

$\exists U, \forall x \in S, x \le U.$

Any number U with this property is said to be an upper bound for S.　●

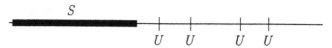

Fig 5.1 Each of the numbers marked U is an upper bound for S.

As an illustration, consider the set S consisting of all the numbers of the form $3 - n$ where $n \in \mathbb{N}$. Each of the numbers $U = 2, 7, \pi$ is an example of an upper bound for this set.

● *Definition 2*

A set S of numbers is said to be bounded below if there is a number which is less than all members of S. Symbolically

$\exists L, \forall x \in S, x \ge L.$

Any number L with this property is said to be a lower bound for S.　●

Fig 5.2 Each of the numbers marked L is a lower bound for S.

As an illustration, consider the set S consisting of all the numbers of the form $n + 1/n$ where $n \in \mathbb{N}$. Each of the numbers $L = -3, 0, 2$ is an example of a lower bound for this set.

● *Definition 3*

A set S of numbers is said to be bounded if it is bounded above and also bounded below. Symbolically

$\exists L, \exists U, \forall x \in S, L \le x \le U.$　●

Fig 5.3 *S* is bounded. Note that *S* can have several components.

As an illustration, consider the set *S* consisting of all the numbers of the form $1 - 1/n$ where $n \in \mathbb{N}$. Each of the numbers $U = 1, 2, \pi$ is an example of an upper bound for this set. Each of the numbers $L = -3, -2\pi, 0$ is an example of a lower bound for *S*.

Definition 3 involves two quantified variables *L* and *U*. The next result shows that these can be replaced by just one.

● *Proposition 2*

A set S is bounded if and only if $\exists M, \forall x \in S, |x| \le M$. ●

PROOF
Now $|x| \le M$ is equivalent to $-M \le x \le M$. So $L = -M$ and $U = M$ satisfy Definition 3. This shows that if the set *S* has the property in the statement of the proposition, then it has the property in Definition 3. To complete the solution we must establish the converse, namely that if *S* satisfies Definition 3 then *S* has the property given in this proposition. Every member *x* of the set *S* satisfies $L \le x \le U$. We let *M* denote the larger of the two numbers $|L|, |U|$. This means that $|L| \le M$ and $|U| \le M$, so that $-M \le L \le M$ and $-M \le U \le M$. Putting together all these inequalities gives $-M \le L \le x \le U \le M$, i.e. $|x| \le M$. ●

If we look at Fig 5.3 we can see that we should be able to move an upper bound *U* to the left until it meets the set *S*, i.e. that there should be a smallest upper bound for *S*. We firstly remark that if $x \in S$ then no number less that *x* can be an upper bound for *S*. So the set of all upper bounds for *S* is bounded below.

● *Definition 4*

If *S* is a set which is bounded above then a number *l* is said to be the least upper bound (l.u.b.) for *S* if

(i) *l* is an upper bound for *S*,
(ii) there is no number less than *l* which is also an upper bound for *S*, i.e. given any
    number less than *l*, we can find a larger member of *S*. ●

We can express these properties symbolically:

(i) $\forall x \in S, x \le l$,
(ii) $\forall \beta > 0, \exists x \in S, x > l - \beta$.

We can illustrate this with a diagram on a number line.

Fig 5.4 The positive number β can be as small as we like, and *x* depends on β.

TUTORIAL PROBLEM 2

(i) You will notice that Definition 4 talks about *the* l.u.b. rather than *a* l.u.b. There is an implicit assumption of uniqueness. Discuss how it follows from the definition that a set cannot have more than one l.u.b.

(ii) Formulate an appropriate definition for greatest lower bound (g.l.b.). Express it in words, in symbols, and illustrate it on a number line.

In several texts the word *supremum* (*sup.*) is used as a synonym for least upper bound, and *infimum* (*inf.*) for greatest lower bound.

One thing which our pictures have not made clear is whether the least upper bound belongs to the set. This may or may not be the case, and a couple of examples will illustrate this. The intuition is clear when they are imagined on a number line, but we shall show that they fit with the more abstract symbolic definition in each case.

Example 1

Let S denote the set of all rational numbers r satisfying $r \leq 3$. Show that 3 is the l.u.b. of S. (In this case $3 \in S$ so S includes its l.u.b.)

From the given inequality $r \leq 3$ so that part (i) of Definition 4 is satisfied. Now whatever the value of $\beta > 0$, $3 > l - \beta$ and $3 \in S$ so that part (ii) of the definition is satisfied. Hence 3 is the l.u.b. for S.

Example 2

Let T denote the set of all rational numbers r satisfying $r < 2$. Show that 2 is the l.u.b. of S. In this case $2 \notin T$ and so T does not include its l.u.b.

Part (i) is demonstrated as in the last example. For all $r \in T$, $r < 2$ so that 2 is an upper bound for T. The second part is a little less straightforward since $2 \notin T$. It relies on the result that between any two numbers it is always possible to find a rational number. This is proved in Proposition 7 below. This means that for any value of $\beta > 0$ it is possible to find a rational number r satisfying $2 - \beta < r < 2$. Since $r < 2$, $r \in T$ and so part (ii) is verified.

One's intuition suggests that the least upper bound and greatest lower bound correspond to the largest and smallest members of the set. When these numbers exist this is indeed the case, but a bounded set need have no largest or smallest member, as the following examples illustrate.

Example 3

Let S denote the set of rational numbers satisfying $x < 6$. Then S has no largest member.

The number 6 is not the largest member of S because it does not belong to S, and indeed no number greater than 6 belongs to S. On the other hand if l is a member of S which is less than 6 then the number $(l + 6)/2$ is a rational number which is greater than l, and less than 6, so that it is a member of S. So l could not be the largest member of S. This demonstrates that S has no largest member.

Of rather more interest is the following example, relating to the previous discussion concerning numbers whose square is equal to 2.

Example 4

Let S denote the set of all rational numbers whose squares are less than 2. Then S is a bounded set but it has no largest member.

If $x \geq 2$ then $x^2 \geq 4 > 2$, and so $x \notin S$. Hence, the number 2 is an upper bound for S. It is clearly not the least upper bound, as a similar argument demonstrates that $3/2$, for example, is also an upper bound for S. Similarly, -2 is a lower bound for S, so that S is a bounded set. Suppose that l were the largest member of S. We cannot have $l^2 = 2$, for there is no such rational number, as we have shown. We cannot have $l^2 > 2$ for then $l \notin S$. We shall show that $l^2 < 2$ is also impossible.

If $l^2 < 2$ we shall show that we can find a larger number $l + \alpha$ ($\alpha > 0$) whose square is also less than 2, as follows:

$$(l + \alpha)^2 = l^2 + 2l\alpha + \alpha^2$$
$$< l^2 + 4\alpha + \alpha^2 \quad \text{since} \quad l < 2$$
$$< l^2 + 5\alpha \quad \text{provided} \quad \alpha < 1$$
$$< 2 \quad \text{provided} \quad \alpha < (2 - l^2)/5.$$

Proposition 7 below ensures that the value of α satisfying the last inequality can be chosen to be a positive rational number.

The final example in this section is a somewhat more abstract one to show how the definition can be used in general circumstances.

Example 5

Let A and B be sets of numbers which are bounded above, with l.u.b. a and b respectively. Let C denote the set of all numbers of the form $x + y$, where $x \in A$ and $y \in B$. Show that the l.u.b. of C is $a + b$.

For all $x \in A$ and $y \in B$, $x \leq a$ and $y \leq b$, using part (i) of the definition of l.u.b. Hence, $x + y \leq a + b$, and so $a + b$ is an upper bound for C. We now apply part (ii) of the definition of l.u.b. to A and to B, with β replaced by $\beta/2$. (You may need some tutorial discussion to explore the logic of this.) So for any positive number β, there are numbers $x \in A$ and $y \in B$ satisfying $x > a - \beta/2$ and $y > b - \beta/2$. Thus, $x + y > a + b - \beta$ and so we have shown that there is a number in C exceeding $a + b - \beta$. So part (ii) of the definition is satisfied, showing that $a + b$ is the l.u.b. for C.

In §5.2 we discussed in outline some of the historical developments leading to the idea among mathematicians that it was necessary to have a precise description of a number which did not depend upon intuitions relating to the number line. There are, broadly speaking, two approaches to this. One is to take the rational number system as given in a precisely described form and use this to *construct* a set which has all the properties we wish the real numbers to have. This is the approach which Dedekind and Cantor adopted. The other approach is to develop the real numbers as an axiomatic system in the same way that Peano did for the integers, and this is the method we shall adopt here. We clearly want the real numbers to form an extension of the rational numbers, so we shall take as axioms those rules which the rationals obey. We must then have an axiom which will remedy the 'deficiency' of the rationals in having no number whose square is equal to 2 (and many other such properties). In terms of a number line we do not want there to be a 'hole' where such a number ought to be. The rational number system is incomplete in this sense and the final axiom remedies this.

● The axiom of completeness for the real number system

This states that every non-empty set of real numbers which is bounded above must have a least upper bound in the set of real numbers. ●

Let us immediately use this axiom to show that there is a real number whose square is equal to 2.

● Proposition 3

There is a real number satisfying the equation $x^2 = 2$. ●

PROOF
We consider the set of all rational numbers r satisfying $r^2 < 2$. We know that this set is bounded above, and the axiom of completeness says that S must have a least upper bound l. We shall show that $l^2 = 2$ by demonstrating that $l^2 < 2$ and $l^2 > 2$ are both impossible. We can make use of the calculations in Example 4. Firstly, we showed there that if $l^2 < 2$ then we can find a larger number $l + \alpha$ whose square is also less than 2, implying that l could not be an upper bound for S. Now suppose that $l^2 > 2$. We shall show that there is a number of the form $l - \beta$ where $\beta > 0$ whose square is also greater than 2. This will mean that $l - \beta$ is also an upper bound, so that l could not be the *least* upper bound. Now

$$(l - \beta)^2 = l^2 - 2l\beta + \beta^2$$
$$> l^2 - 4\beta \quad \text{(since } \beta^2 > 0 \text{ and } l < 2\text{)}$$
$$> 2 \quad \text{provided} \quad \beta < (l^2 - 2)/4.$$

So we have shown that there is a real number whose square is 2, finally legitimizing the use of the symbol $\sqrt{2}$ for the number l defined above. ●

This axiom will also guarantee the existence of the solution of many equations besides $x^2 = 2$. From elementary mathematics we have the formula for the solution of a quadratic equation. Sometimes there will not be real solutions (e.g. $x^2 + 1 = 0$) but where there are they all involve square roots. Clearly we can envisage more complicated polynomial equations such as $x^7 - 5x^4 + 2x^3 - x + 3 = 0$. Numbers which satisfy such equations are called algebraic numbers. One of the remarkable things about the real number system is that the set of all these algebraic numbers together with the rational numbers forms only a tiny part of the real number system. Cantor himself was able to demonstrate that, in a certain sense, the set of numbers which are not algebraic is infinitely larger than the set of those which are. So what do these other numbers (called transcendental numbers) look like. Well, curiously, we do not have names or symbols for most of them. Some common numbers like π and the exponential number e have been shown to be transcendental, but a number like $\pi + e$ has so far defied all attempts to determine whether or not it is transcendental. This, in fact, makes us think what on earth we might mean by the sum of two numbers about which we know very little. Certainly their decimal expansions are in no way regular so we cannot use that as an approach to defining the sum of two real numbers. Such is the power of the axiom of completeness that we can use this to add any two real numbers, assuming only that we know how to add rationals. This is discussed in the next section of this chapter.

EXERCISES 5.3

1. For each of the sets described below, say whether it is bounded above, and/or bounded below, and if so what its least upper bound and/or greatest lower bound are. State also whether the set has a largest and/or smallest member, and if so identify it.

    (i) The set of all even positive integers.

    (ii) The set of all rational numbers r satisfying $0 \le r < 1$.

    (iii) The set of all values of x for which $x = \sin t$ for some real number t.

    (iv) The set of all numbers of the form $3^{-m} + 5^{-n}$, where m and n are any positive integers.

    (v) The set of all real numbers x satisfying $-1 \le \tan x \le 1$.

    (vi) The set of numbers of the form $1 + 1/n$, where n is any positive integer.

    (vii) The set of positive integers n satisfying $n^2 \le 10$.

    (viii) The set of real numbers y of the form $y = (2x + 5)/(x + 1)$, where x can be any positive real number.

2. Let S denote the set of rational numbers whose squares are less than 8. Show that S is bounded and that it has no largest member. Show that S has no smallest member.

3. Let S denote a set of numbers which is bounded above, with l.u.b. l. Let T denote the set of all numbers of the form $-x$, where x can be any member of S. Show, using the definition, that T is bounded below and that g.l.b.$T = -$l.u.b.S.

4. Let A and B be bounded sets of numbers. Let D denote the set of all numbers of the form $x - y$, where $x \in A$ and $y \in B$. Find an example using intervals (sets of numbers of the form $p \le x \le q$) to show that it is not true in general that

   $$\text{l.u.b.}\,D = \text{l.u.b.}\,A - \text{l.u.b.}\,B.$$

   Explore some more examples and try to find some general relationships which are valid, involving the l.u.b. and g.l.b. of A, B and D.

5. Investigate the situation similar to that of Exercise 4, with subtraction replaced by multiplication.

6. Let S be a set of numbers which is bounded above. Let T denote the set of all upper bounds of S. Show that T is bounded below, that it contains its g.l.b., and that $\text{g.l.b.}\,T = \text{l.u.b.}\,S$.

5.4 Arithmetic and Algebra with Real Numbers

The definition of real numbers using the axiom of completeness is a rather abstract one, and so the question arises as to how we might add and multiply real numbers like $\sqrt{2}$. This is quite different from the situation with integers and rational numbers. In the latter case we have the familiar rule for adding fractions discussed in Chapter 3. To develop the whole of the theory of arithmetic for real numbers includes defining addition and multiplication as well as the relation of ordering, establishing that all the familiar algebraic properties are satisfied, for example that $a(b + c) = ab + ac$, and demonstrating consistency with the existing arithmetic for rational numbers. This is an enterprise which is outside the scope of this book, but we shall give a flavour of what is involved by outlining the method of defining addition of two real numbers.

Let h and k denote any two real numbers. We let S denote the set of rational numbers a less than h, and T denote the set of rational numbers b less than k. We now define the set U to be the set of all rational numbers of the form $a + b$, where $a \in S$ and $b \in T$. Since S and T are bounded above there are rational numbers p and q which are greater than h and k respectively. Since $a < p$ and $b < q$ we have $a + b < p + q$, where the operation of addition is being performed between rational numbers. Thus, the set U is bounded above, with $p + q$ as an upper bound. The axiom of completeness tells us that this set U therefore has a least upper bound, and we define this least upper bound to be the sum $h + k$.

As an example of the necessity of establishing that real numbers satisfy the rules of arithmetic, we shall give an outline of what is necessary in order to show that $\sqrt{2} \times \sqrt{3} = \sqrt{6}$.

● *Proposition 4*

There is a unique positive real number satisfying the equation $x^2 = 2$. ●

PROOF

Existence was established in Proposition 3. Here we are demonstrating uniqueness. The proof uses the method of contradiction. Suppose that x_1 and x_2 are both positive numbers satisfying $x^2 = 2$. Then $x_1^2 = x_2^2$, and so $x_1^2 - x_2^2 = 0$. Factorizing this gives $(x_1 + x_2)(x_1 - x_2) = 0$. We must therefore have $(x_1 + x_2) = 0$ or $(x_1 - x_2) = 0$. Since x_1 and x_2 are both positive we cannot have $(x_1 + x_2) = 0$. Hence $(x_1 - x_2) = 0$ and so $x_1 = x_2$, proving uniqueness. ●

TUTORIAL PROBLEM 3

Find all the places in the above proof where assumptions have been made about the algebraic properties of real numbers, and list these properties.

● *Proposition 5*

$\sqrt{2} \times \sqrt{3} = \sqrt{6}$ ●

PROOF

Let $a = \sqrt{2}$, $b = \sqrt{3}$ and $c = \sqrt{6}$. In other words, let a, b, c denote the unique positive real numbers satisfying $x^2 = 2$, $x^2 = 3$ and $x^2 = 6$ respectively, whose existence relies on the axiom of completeness. So $a^2 = 2$ and $b^2 = 3$. Multiplying these two equations together gives $a^2 \times b^2 = 2 \times 3 = 6$. The associative law of multiplication for real numbers enables us to regroup the left-hand side of this equation to give $(a \times b)^2 = 6$. Now a and b are positive and so $a \times b$ is positive. (Here is another assumption which would need to be justified in a complete account.) But there is only one positive number, c, satisfying $x^2 = 6$, so that $a \times b = c$, i.e. $\sqrt{2} \times \sqrt{3} = \sqrt{6}$. ●

We shall now consider some relationships between irrational numbers and rationals. We first show that they are thoroughly intermixed on the number line, that is, we cannot find any intervals consisting entirely of rationals or entirely of irrationals.

● *Proposition 6*

Given any two positive real numbers a and b, there is a positive integer n satisfying $na \geq b$. (This is known as the Archimedean Property.) ●

PROOF

As on so many occasions, the proof employs the method of contradiction. So we suppose that the result is false, i.e. that there are two real numbers a and b such that for every positive integer n, $na < b$. This means that the set S of numbers of the form na is bounded above by b. S therefore has a l.u.b. l. Now $l - a < l$ and so there is a member of the set S between $l - a$ and l, i.e. there is a positive integer n_0 satisfying $l - a < n_0 a < l$. Adding a to each side of the left-hand inequality gives $l < (n_0 + 1)a$, giving a member of S greater than l. This is a contradiction. ●

● *Proposition 7*

Between any two real numbers there are both rational and irrational numbers. ●

PROOF

We shall prove the result for positive numbers. Let p and q denote two positive real numbers with $p < q$. The method of proof essentially looks at the gap between p and q, finds a number of the form $1/n$ smaller than this gap, where n is a positive integer, and then counts along in units of $1/n$. It is impossible to jump over the gap with this unit, and so there will be a multiple of $1/n$ in the gap. We can formalize this as follows.

Since $0 < q - p$, we have $0 < 1/(q - p)$. We use the Archimedean Property with $a = 1$ and $b = 1/(q - p)$. Consequently, there is a positive integer n satisfying $n > 1/(q - p)$. Rearranging this inequality gives $0 < (1/n) < q - p$. Again using the Archimedean Property gives a positive integer m satisfying $m(1/n) \geq q$. Let m_0 denote the smallest such m. Then

$$\frac{m_0 - 1}{n} < q \quad \text{and} \quad \frac{m_0}{n} \geq q.$$

Putting these inequalities together gives

$$\frac{m_0 - 1}{n} = \frac{m_0}{n} - \frac{1}{n} \geq q - \frac{1}{n} > q - (q - p) = p.$$

We have therefore shown that

$$p < \frac{m_0 - 1}{n} < q,$$

giving a rational number between p and q. Furthermore we have

$$0 < \frac{1}{n\sqrt{2}} < \frac{1}{n} < (b - a),$$

and by the process above we can find a positive integer m_1 satisfying

$$p < \frac{m_1}{n\sqrt{2}} < q,$$

giving an irrational number between p and q. ●

TUTORIAL PROBLEM 4

Fill in the details of the second part of the above proof. Explain how to extend the result to the case where p and q are not both positive.

We now look at an example of an algebraic relationship involving rational and irrational numbers.

Example 6

Suppose that a and b are rational numbers satisfying $a + b\sqrt{2} = 0$. Prove that $a = b = 0$.

If $b \neq 0$ then we can rearrange this equation to give $\sqrt{2} = -a/b$. But $-a/b$ is a rational number. This is a contradiction since $\sqrt{2}$ is irrational. So we must have $b = 0$, and therefore $a = 0$.

TUTORIAL PROBLEM 5

Let us now generalize this a little and consider the equation $a + b\sqrt{2} + c\sqrt{3} = 0$, where a, b, c are rational numbers. Rearrange the equation by subtracting a from both sides. Squaring both sides of the new equation will give a relationship similar to that of Example 6 but involving $\sqrt{6}$. Use this to show that we must have $a = b = c = 0$.

EXERCISES 5.4

1. Find a rational number between $7/2$ and $\sqrt{2}$.
2. Find an irrational number between $2/3$ and $3/4$.
3. Decide whether each of the following is always true, sometimes true, or always false:
   (i) rational $\times$ rational $=$ rational,
   (ii) irrational $+$ irrational $=$ irrational,
   (iii) rational $+$ irrational $=$ irrational,
   (iv) rational $\times$ irrational $=$ irrational,
   (v) irrational $\times$ irrational $=$ irrational.

Summary

The topics discussed in this chapter are among the most conceptually demanding in basic mathematics. The brief historical sketch in §5.2 indicates the difficulties involved in developing the ideas to the form we have them today. We mentioned that a well-developed theory of real numbers is necessary for a theory of limits, and we shall look at this in relation to sequences in Chapter 7. The ideas are also vital for a proper foundation for the theory of functions and calculus, and are discussed in a sequel to this book, on Analysis.

EXERCISES ON CHAPTER 5

1. For each of the sets described below, say whether it is bounded above, and/or bounded below, and if so what its least upper bound and/or greatest lower

bound are. State also whether the set has a largest and/or smallest member, and if so identify it.

(i) The set of all prime numbers.

(ii) The set of numbers x of the form $x = s + t$ where $-1 \leq s < 1$ and $-1 < t \leq 1$.

(iii) The set of numbers of the form $(-1)^n n$, where n is any positive integer.

(iv) The set of rational numbers x satisfying $x^2 < x$.

2. Let A and B denote bounded sets of positive numbers. Let C denote the set of all numbers of the form $a \times b$, where $a \in A$ and $b \in B$. Prove from the definition that

$$\text{l.u.b.}C = \text{l.u.b.}A \times \text{l.u.b.}B.$$

3. Let A and B denote bounded sets of positive numbers. Let D denote the set of all numbers of the form $a \div b$, where $a \in A$ and $b \in B$. Find examples to show that it is not necessarily true that

$$\text{l.u.b.}C = \text{l.u.b.}A \div \text{l.u.b.}B.$$

4. Show that if $a + b\sqrt[3]{2} + c\sqrt[3]{4} = 0$, where a, b, c are rational numbers, then $a = b = c = 0$.

5. Explain how the axiom of completeness together with the result of Exercise 5.3(3) implies that every non-empty set of real numbers, which is bounded below, has a greatest lower bound in the set of real numbers.

6. Show that there is a real number satisfying the equation $x^3 = 4$. Prove that it is unique.

6 • Complex Numbers

In previous chapters we have discussed extensions of the number system, one important aim being to increase the kinds of equations which can be solved. So, if we start with the natural numbers, in order to be able to solve equations such as $4 + x = 2$ we have to create the integers. To deal with $2x = 3$ we need to construct the rational numbers. Finally, we analysed what is needed to characterize the real number system to solve equations like $x^2 = 2$. There are still equations with no solution however, for example $x^2 + 1 = 0$, and it is partly in response to this that we extend the number system once again. A formal solution of $x^2 + 1 = 0$ would be $x = \sqrt{-1}$, but we know from Axiom O4 and Exercise 1 of §4.1 that the square of any non-zero real number is positive and so cannot equal -1.

It is interesting historically that one of the early appearances of $\sqrt{-1}$ was not in connection with quadratic equations but with cubics. The solution of quadratics had been known for a very long time, and it would have been said that $ax^2 + bx + c = 0$ had no roots when $b^2 - 4ac < 0$. With cubics the situation was different. In the 16th century, Italian algebraists discovered how to solve equations of degree 3 and 4. The chief figures involved were Bombelli, Scipione del Ferro, Cardano, Tartaglia and Vieta. It is interesting to read about the mathematics they used, as well as the chicanery and intrigue involved in the various efforts to claim precedence for the invention of the method. This method gave rise to square roots of negative numbers even when the cubic had all three roots real. An example, which was used in the writings of the time, considered the equation $x^3 - 15x - 4 = 0$. One obvious solution of this is $x = 4$, and it is then straightforward to show that the other two solutions are $x = -2 \pm \sqrt{3}$. The general formulae developed gave rise to the expression

$$x = \sqrt[3]{2 + \sqrt{-121}} + \sqrt[3]{2 - \sqrt{-121}}.$$

Expressions such as this seem to have been used in a purely formal sense, without any meaning being attached to them. Words such as 'sophistry' appear, with the comment that they are 'as subtle as they are useless'. Cardano called them *numeri ficti*, and the fact that we still speak of *imaginary numbers* today is testimony to the persistence of such a philosophical attitude. Such numbers occurred from time to time thereafter, and luminaries such as Descartes and Newton had doubts about them. Another occurrence involved the attempt to integrate the reciprocal of a quadratic with no real roots by the method of partial fractions. This led to the appearance of logarithms of imaginary numbers, a matter of controversy between Leibnitz and J. Bernoulli. Euler wrote on the matter in a famous publication of 1749: *De la controverse entre Mrs. Leibnitz & Bernoulli sur les Logarithmes des Nombres Negatifs et Imaginaires.* (Notice the 18th century abbreviation for Messieurs.)

Applications of these new numbers were considered by the great German mathematician, physicist and astronomer Carl Friedrich Gauss (1777–1855). It is clear, through the process of completing the square, that if we can solve $x^2 + 1 = 0$ then we can solve any quadratic equation, and the investigations alluded to above indicated that cubic and quartic equations would also be soluble. But what about equations containing higher powers of x? Would we need to extend the number system yet again in order to bring these equations within the scope of numeric solution? What Gauss was able to prove in his doctoral dissertation in 1799 was that the complex numbers provide the roots of all polynomial equations. We note in passing that he did not do this by generalizing the formulae for cubics and quartics, indeed the work of Galois and Abel was to show that in an algebraic sense there could be no general formulae for equations of degree higher than 4.

The philsophical question of what the square root of a negative number could possibly mean was answered to some extent by showing that it could be represented geometrically. Like the algebraic development, this took some time, with a number of mathematicians generating the essential ideas, including Descartes (1637), Wallis (1673), Wessel (1797), Argand (1806) and Gauss himself, who published an account of the representation of complex numbers in the plane in 1831. He had been aware of this some 20 years earlier, but it seems that his natural caution led him to be wary of publicly acknowledging ideas which were still felt to be philosophically suspect.

It was the Irish mathematician, linguist and astronomer William Rowan Hamilton (1805–1865) who finally showed in 1833 how one could define complex numbers solely in terms of real numbers without the need for a fictitious quantity like a non-existent square root of a negative number. The story of this and his subsequent investigations is a fascinating one, culminating in his invention of a hypercomplex number system, the *quaternions*, and the famous (if not apocryphal) episode of his scratching the crucial equations on a stone bridge. Hamilton's approach is at root the one we adopted in §3.2, that of *constructing* one system from another. We shall explore this in some detail in the following section.

The story does not end there. A great deal has been done during the past two centuries in calculus and analysis involving complex numbers and functions, giving rise to one of the most powerful and satisfying branches of mathematics, which is a cornerstone of many undergraduate courses.

6.1 Hamilton's Definition

A common approach to the introduction of complex numbers is to define them as 'expressions of the form $a + bi$ where a and b are real numbers and $i = \sqrt{-1}$'. We then multiply them as if they obeyed all the rules of algebra to obtain

$$(a + bi)(c + di) = ac + bci + adi + bdi^2 = (ac - bd) + (ad + bc)i.$$

Unfortunately this does not tell us what $\sqrt{-1}$ means. Replacing '$i = \sqrt{-1}$' by '$i^2 = -1$' does not help, because we have nothing to tell us that there is a solution to this equation. The symbol 'i' remains mysterious and imaginary. Modern algebra has the notion of a field extension by adjoining an undefined variable satisfying specified relations, but this is far too abstract for an introductory approach.

The solution to the problem arises from the work of Hamilton in 1833 referred to in the introduction. He gave a formal definition of complex numbers in terms of ordered pairs. The fact that the geometrical representations involved coordinates fitted in very well with the ordered pair idea. We shall give his definition and look at some of the consequences, in particular showing how it leads to the traditional notation for complex numbers used at the beginning of this section.

● Definition I

The set $\mathbb{C}$ of complex numbers consists of all ordered pairs (a, b) of real numbers, with operations of addition and multiplication defined by

$$(a, b) + (c, d) = (a + c, b + d),$$
$$(a, b) \times (c, d) = (ac - bd, ad + bc).$$
●

Naturally, Hamilton formulated his definitions of the arithmetic operations to reflect the way the traditional notation worked. He certainly did not invent arbitrarily what at first sight appears to be a bizarre way of describing multiplication.

We showed in §3.2 that the set of rational numbers contains a subset which behaves like the integers. We can show here that in this sense the real numbers are embedded in the complex numbers, for if we consider the subset of $\mathbb{C}$ for which the second component is zero we have

$$(a, 0) + (c, 0) = (a + c, 0) \qquad \text{and} \qquad (a, 0) \times (c, 0) = (ac, 0).$$

So this subset behaves exactly like the real numbers in respect of arithmetic and algebra.

This means that the complex number $(-1, 0)$ is the image of the real number -1. Furthermore the definition of multiplication tells us that $(0, 1) \times (0, 1) = (-1, 0)$. It is this relationship which gives meaning to $\sqrt{-1}$ in terms of real numbers, for we have shown that $(0, 1)$ is an ordered pair of real numbers which when multiplied by itself gives the embedded image of -1.

Now that Hamilton's definition has given a meaning to and established the mathematical existence of complex numbers in terms of real numbers, we shall make the link with the traditional notation. Given a complex number (a, b), it is straightforward to verify that the definitions of addition and multiplication give

$$(a, b) = (a, 0) + (b, 0) \times (0, 1).$$

We have seen that numbers with zero as the second component mimic the behaviour of real numbers, and so we can legitimately abbreviate $(a, 0)$ and $(b, 0)$ as a and b. We shall then abbreviate $(0, 1)$ with the symbol 'i', so that we can now write

$(a, b) = a + bi$, where we have adopted the usual procedure of denoting multiplication by juxtaposition, i.e. writing $b \times i$ simply as bi.

Does -1 have any other square roots? The implication $x^2 = y^2 \Rightarrow x = \pm y$ uses only the algebraic properties of the number systems for its justification, and it applies to the complex numbers as well as to the integers, rationals and reals. So every non-zero complex number has exactly two square roots. In the case of -1 these are i and $-i$.

6.2 The Algebra of Complex Numbers

As far as solving equations is concerned, the algebraic rules are the same as those of the rational numbers as given in Chapter 3, as we shall show. This serves as a general demonstration of what is involved in verifying that a given system obeys a particular set of axioms. We use the Hamilton notation in the following proposition as an illustration that it can be used in calculations as well as simply in definitions. Thereafter, we shall revert to the traditional notation for complex numbers.

● *Proposition I*

The complex numbers obey the axioms for a field. ●

PROOF

We refer to the axioms by the labels used in Chapter 3.

A1. This follows from the definition of addition (Definition 1 above).

A2. This is inherited from the corresponding property for the real numbers. Using the Hamilton notation we have

$$\begin{aligned}
(a, b) + ((c, d) + (e, f)) &= (a + (c + e), b + (d + f)) \quad \text{from the definition,} \\
&= ((a + c) + e, (b + d) + f) \quad \text{using associativity in R,} \\
&= ((a, b) + (c, d)) + (e, f) \quad \text{from the definition.}
\end{aligned}$$

A3. For all $(a, b) \in \mathbb{C}$, $(a, b) + (0, 0) = (a, b)$ from the definition of addition. So $(0, 0)$ is the additive identity in $\mathbb{C}$. It corresponds of course to the real number 0.

A4. For all $(a, b) \in \mathbb{C}$, $(a, b) + (-a, -b) = (0, 0)$ from the definition of addition. So $(-a, -b)$ is the additive inverse of (a, b).

A5. This is inherited from the corresponding property for the real numbers, giving

$$(a, b) + (c, d) = (a + c, b + d) = (c + a, d + b) = (c, d) + (a, b).$$

M1. This follows from the definition of multiplication (Definition 1 above).

M2. This involves some tedious algebra. We commented on this in Exercise 2 of §3.2. The calculations are as follows. Observe that they use distributivity and associativity for the real numbers.

$$(a, b) \times ((c, d) \times (e, f))$$
$$= (a, b) \times (ce - df, cf + de)$$
$$= (a(ce - df) - b(cf + de), a(cf + de) + b(ce - df))$$
$$= (ace - adf - bcf - bde, acf + ade + bce - bdf)$$
$$= (ace - bde - adf - bcf, acf - bdf + ade + bce)$$
$$= ((ac - bd)e - (ad + bc)f, (ac - bd)f + (ad + bc)e)$$
$$= (ac - bd, ad + bc) \times (e, f)$$
$$= ((a, b) \times (c, d)) \times (e, f).$$

M3. For all $(a, b) \in \mathbb{C}, (a, b) \times (1, 0) = (a, b)$. So $(1, 0)$ is the multiplicative identity, corresponding to the real number 1.

M4. Here we have to try to solve the equation $(a, b) \times (x, y) = (1, 0)$. Using the definition of multiplication this gives $(ax - by, ay + bx) = (1, 0)$. Two ordered pairs are equal if and only if the respective components are equal. This leads to the simultaneous linear equations

$$ax - by = 1,$$
$$ay + bx = 0.$$

It is left to the reader to verify that, provided $a^2 + b^2 \neq 0$, the solution of these equations is

$$x = \frac{a}{a^2 + b^2}, \quad y = -\frac{b}{a^2 + b^2}.$$

So every non-zero complex number has a multiplicative inverse. For real numbers we also refer to this as the reciprocal, and we shall do so for complex numbers. We shall also use the normal reciprocal notation in the case of complex numbers. The result here can therefore be written in the traditional notation as

$$\frac{1}{a + bi} = \frac{a}{a^2 + b^2} - \frac{b}{a^2 + b^2} i.$$

M5. We use commutativity for multiplication of real numbers, so that

$$(a, b) \times (c, d) = (ac - bd, ad + bc)$$
$$= (ca - db, da + cb)$$
$$= (c, d) \times (a, b).$$

The final axiom is that of distributivity, and we shall leave that as an exercise. ●

TUTORIAL PROBLEM I

Undertake a careful analysis of the proof of M2 above and decide exactly which properties of addition and multiplication in the real numbers have been used at each stage.

The next set of axioms we have to consider is that of order. This is dealt with as follows.

● Proposition 2

It is impossible to define an order relation on $\mathbb{C}$ which is compatible with addition and multiplication. ●

PROOF

Compatibility with addition and multiplication refers to Axioms O3 and O4 in §4.1, which govern the manipulation of inequalities. We shall show that the assumption of the existence of a relation of inequality satisfying the order axioms leads to a contradiction. Now Axiom O4 and Example 1 of §4.1 demonstrate that the square of any non-zero number is greater than zero. This used only axioms from the list, and so applies to the complex numbers as well as to the rational and real number systems. So $-1 = i^2 > 0$, and adding 1 to both sides of this inequality by Axiom O3 gives $0 > 1$. But 1 is a square $(1 = 1^2)$ and so $1 > 0$. This contradicts Axiom O1. So there is no order relation on $\mathbb{C}$ which satisfies the axioms. This means that it makes no sense to talk about one complex number being larger or smaller than another. ●

Having discussed the algebraic structure of $\mathbb{C}$ we shall now consider some of the quantities used in the arithmetic and geometry of complex numbers. It is traditional to use z to denote an arbitrary complex number, analogous to the use of x for real numbers.

● Definition 2

Given a complex number $z = x + yi$ we define the following quantities.

(i) x is called the real part of z, and is denoted by $\mathrm{Re}(z)$.

(ii) y is called the imaginary part of z, and is denoted by $\mathrm{Im}(z)$.

(iii) The complex conjugate of z is defined to be $x - yi$. There are various notations in use, among the most common of which are z^* and $\bar{z}$. We shall use z^*.

(iv) The modulus of z is defined to be the non-negative real number $\sqrt{(x^2 + y^2)}$. It is denoted by $|z|$.

Note that y, the imaginary part of z, is a real number, a point which sometimes causes confusion. This and the other parts of Definition 2 will make more sense in the next section when we shall interpret them geometrically. ●

● Example 1

Show that (i) $z + z^* = 2\,\mathrm{Re}(z)$, (ii) $z - z^* = 2i\,\mathrm{Im}(z)$, (iii) $zz^* = |z|^2$.

This is an exercise in using the definitions.

(i) Adding z to its conjugate gives $z + z^* = (x + yi) + (x - yi) = 2x = 2\,\mathrm{Re}(z)$.

(ii) Subtracting gives $z - z^* = (x + yi) - (x - yi) = 2yi = 2i\,\mathrm{Im}(z)$.

(iii) Multiplying z by its conjugate gives

$$zz^* = (x + yi)(x - yi) = (x^2 + y^2) + (-xy + yx)i = (x^2 + y^2) = |z|^2.$$

In the discussion of Axiom M4 above we found an expression for the reciprocal of a complex number. That expression is consistent with the result of Example 1, which can be written as

$$\frac{1}{z} = \frac{z^*}{|z|^2} \quad (z \neq 0).$$

Example 2

Find the real and imaginary parts of the complex number $\dfrac{3 - 2i}{1 + 3i}$.

The result of Example 1 shows that if we multiply the denominator by its conjugate we obtain a real number. The strategy for this problem is therefore to multiply numerator and denominator of the given number by the conjugate of the denominator. This gives

$$\frac{3 - 2i}{1 + 3i} = \frac{(3 - 2i)(1 - 3i)}{(1 + 3i)(1 - 3i)} = \frac{(3 - 6) + (-9 - 2)i}{1^2 + 3^2} = -\frac{3}{10} - \frac{11}{10}i.$$

So the real part is $-3/10$ and the imaginary part is $-11/10$.

• Proposition 3

Complex conjugation commutes with addition, multiplication and division, i.e.

(i) $(z_1 + z_2)^* = z_1^* + z_2^*$,   (ii) $(z_1 z_2)^* = z_1^* z_2^*$,   (iii) $(1/z)^* = 1/(z^*)$.

PROOF
A comment on the language used here is in order. The use of the word 'commutes' relates to the commutative axioms (A5 and M5 in Chapter 3), which essentially say that the order of operations can be reversed. So the concise phrase 'conjugation commutes with addition' symbolized in (i) means that if we add two complex numbers and then find the conjugate of the sum, we will obtain the same answer by finding the conjugate of each number separately and then adding these conjugate numbers together. The use of the word 'commutes' in this sense occurs a good deal in algebra and geometry.

Proof of the three properties is a matter of algebraic manipulation. We let $z_1 = x_1 + y_1 i$, etc.

(i) $((x_1 + y_1 i) + (x_2 + y_2 i))^* = ((x_1 + x_2) + (y_1 + y_2)i)^*$
    $= ((x_1 + x_2) - (y_1 + y_2)i) = (x_1 - y_1 i) + (x_2 - y_2 i).$

(ii) $((x_1 + y_1 i)(x_2 + y_2 i))^* = ((x_1 x_2 - y_1 y_2) + (x_1 y_2 + y_1 x_2)i)^*$
    $= ((x_1 x_2 - y_1 y_2) - (x_1 y_2 + y_1 x_2)i) = (x_1 - y_1 i)(x_2 - y_2 i).$

(iii) $\left(\dfrac{1}{x + yi}\right)^* = \left(\dfrac{x}{x^2 + y^2} - \dfrac{y}{x^2 + y^2}i\right)^*$

   $= \left(\dfrac{x}{x^2 + y^2} + \dfrac{y}{x^2 + y^2}i\right) = \dfrac{1}{x - yi}.$

● Proposition 4

If the complex number z is a root of a polynomial with real coefficients, then the conjugate z^* is also a root of the polynomial. ●

PROOF

Suppose that $a_n z^n + a_{n-1} z^{n-1} + \ldots + a_2 z^2 + a_1 z + a_0 = 0$, where the coefficients $a_0, \ldots, a_n$ are real numbers.

We first note that if a denotes a real number then $a^* = a$, for $a^* = (a + 0i)^* = a - 0i = a$. Using the results of Proposition 3, extended as in Exercise 3 below, we then have

$$(a_n z^n + a_{n-1} z^{n-1} + \ldots + a_2 z^2 + a_1 z + a_0)^* = 0^* = 0$$
$$a_n^* (z^n)^* + a_{n-1}^* (z^{n-1})^* + \ldots + a_2^* (z^2)^* + a_1^* z^* + a_0^* = 0,$$
$$a_n (z^*)^n + a_{n-1} (z^*)^{n-1} + \ldots + a_2 (z^*)^2 + a_1 z^* + a_0 = 0.$$

Therefore z^* is a root of the polynomial. ●

◉ Example 3

Given that $z = 1 + i$ is a solution of the equation $z^4 - 5z^2 + 10z - 6$, find the other solutions.

The result of proposition 4 tells us that $z = 1 - i$ is also a solution. This tells us that $(z - (1 + i))$ and $(z - (1 - i))$ are both factors of $z^4 - 5z^2 + 10z - 6$, so that $(z - (1 + i))(z - (1 - i)) = z^2 - 2z + 2$ is a factor. Polynomial division then gives

$$z^4 - 5z^2 + 10z - 6 = (z^2 - 2z + 2)(z^2 + 2z - 3) = (z^2 - 2z + 2)(z - 1)(z + 3),$$

so that the four solutions are $1 + i, 1 - i, 1, -3$. ◉

TUTORIAL PROBLEM 2

Discuss and fill in details in the following argument, which generalizes the result of Example 3. Construct examples to illustrate the various assertions.

Suppose we have a polynomial of degree n with real coefficients. We commented in the introduction that Gauss proved that the roots of this are all complex numbers (some of which may be real, i.e. with imaginary part zero). This enables the polynomial to be factorized completely into linear factors (some of which may be repeated). Proposition 4 tells us that non-real roots of such a polynomial occur in conjugate pairs. Corresponding pairs of linear factors will always multiply to give a real quadratic. Thus, every polynomial with real coefficients can be factorized into a product of real linear and real quadratic factors.

In Exercise 5 below you are asked to show that $|z_1 z_2| = |z_1||z_2|$. The analogous result for addition is false, as the following illustration shows. Let $z_1 = 3 + 4i$ and

$z_2 = 5 - 12i$. Then $z_1 + z_2 = 8 - 8i$, and calculating the moduli of these numbers gives $|z_1| = 5$, $|z_2| = 13$ and $|z_1 + z_2| = \sqrt{128} = 8\sqrt{2}$. So $|z_1 + z_2| \neq |z_1| + |z_2|$. The value of the left-hand side is approximately 11.3 whereas the right-hand side is equal to 18. In fact the relative magnitudes are always this way round, as the following result demonstrates.

● Proposition 5—The triangle inequality

For all complex numbers z_1, z_2 we have $|z_1 + z_2| \leq |z_1| + |z_2|$. ●

PROOF
The reason for the name 'triangle inequality' will become clear in the next section, where a geometrical justification will be given. We shall first establish a result to be used in the course of the proof.

Let $z = x + yi$. Then $\text{Re}(z) = x = \sqrt{(x^2)} \leq \sqrt{(x^2 + y^2)} = |z|$. Equality occurs if and only if $y = 0$, i.e. when z is real. We now have

$$
\begin{aligned}
|z_1 + z_2|^2 &= (z_1 + z_2)(z_1 + z_2)^* = (z_1 + z_2)(z_1^* + z_2^*) \\
&= z_1 z_1^* + z_2 z_2^* + z_1 z_2^* + z_1^* z_2 \\
&= |z_1|^2 + |z_2|^2 + z_1 z_2^* + (z_1 z_2^*)^* \\
&= |z_1|^2 + |z_2|^2 + 2\text{Re}(z_1 z_2^*) \\
&\leq |z_1|^2 + |z_2|^2 + 2\left|(z_1 z_2^*)\right| \\
&= |z_1|^2 + |z_2|^2 + 2|z_1||z_2^*| \\
&= |z_1|^2 + |z_2|^2 + 2|z_1||z_2| \\
&= (|z_1| + |z_2|)^2.
\end{aligned}
$$

Taking the square root gives the result. ●

TUTORIAL PROBLEM 3

In the chain of argument in Proposition 5 we have used some results from earlier propositions in this section, parts of Definition 1, and some results from the exercises below. Find the justification for each step in the argument.

EXERCISES 6.2

1. Express the following complex numbers in the form $a + bi$, where a and b are real.

   (i) $(4 - 2i)(-3 + 7i)$, (ii) $\dfrac{1}{7 - 4i}$, (iii) $\dfrac{-1 + 4i}{-4 - 3i}$.

2. Prove from Definition 2(iv) that $|z_1 z_2| = |z_1||z_2|$ and that $|1/z| = 1/|z|$. Extend the first result to n complex numbers by induction.

3. Show from the definition of complex conjugate that $(z^*)^* = z$.

4. Prove that $|z^*| = |z|$.

5. Express $(1 + i)^2$ in the form $a + bi$ where a and b are real. Use this to write down a square root of i. Try to find the other square root of i. Try to find the square roots of $-i$.

6. Solve the quadratic equation $z^2 - 2z - i = 0$.

7. Replace z_1 by $z_1 - z_2$ in the triangle inequality to show that

$$|z_1 - z_2| \geq |z_1| - |z_2|.$$

Deduce that $|z_1 - z_2| \geq ||z_1| - |z_2||$.

6.3 The Geometry of Complex Numbers

We remarked in the introduction that geometrical representations for complex numbers were investigated during the 17th–19th centuries. The Hamilton notation itself is of the same form as that used in two-dimensional coordinate geometry, and so it seems natural to represent a complex number $x + yi$ as the point (x, y) in the Cartesian coordinate plane. This representation is referred to variously as the Argand Diagram, the Gaussian plane etc. Not wishing to single out any one particular mathematician above the others involved, we shall follow the practice of referring simply to the complex plane. In this context, the x and y-axes are called the real axis and the imaginary axis respectively.

Representing complex numbers in the plane would not be especially significant unless the various operations and quantities associated with complex numbers had an interpretation, and we shall explore this here and in subsequent sections of this chapter. Adding a visual component to the algebraic aspect of complex numbers will enable a much richer concept image for complex numbers to be established. Firstly we shall illustrate the basic quantities described in Definition 2.

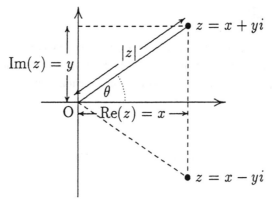

Fig 6.1 The complex plane: real and imaginary parts, conjugate and modulus.

We shall now consider another important quantity associated with the complex number z.

• Definition 3

The argument of a non-zero complex number z is the angle θ between the line joining O to z and the positive real axis. It is denoted by $\arg(z)$. In order to associate a unique such angle with z we adopt the convention that $-\pi < \theta \le \pi$.   ●

The choice of the interval for θ is particularly important in applications in calculus. It is often referred to as the principal argument. The angle is marked on Fig 6.1. The argument can be determined from the real and imaginary parts, because we can see from Fig 6.1 that $\tan\theta = y/x$. However, it is insufficient to try to invert this formula as $\theta = \tan^{-1}(y/x)$. This is because the tangent function has period π. By way of illustration of this point, if we plot the complex numbers $1 + i$ and $-1 - i$ we can see that $\arg(1 + i) = \pi/4$ and $\arg(-1 - i) = -3\pi/4$. However, in both cases the value of y/x is equal to 1, so the quantity y/x cannot be used to determine the argument without considering which quadrant the complex number is in. The value of the inverse tangent is always between $\pm\pi/2$, as most electronic calculators will confirm.

TUTORIAL PROBLEM 4

> Discuss how the following specification determines the value of the argument of the complex number $x + yi$. Give some numerical examples and illustrate them in the complex plane.
>
> $$\arg(x + yi) = \begin{cases} \tan^{-1}\left(\frac{y}{x}\right) & \text{if } x > 0; \\ \frac{\pi}{2} & \text{if } x = 0 \text{ and } y > 0; \\ -\frac{\pi}{2} & \text{if } x = 0 \text{ and } y < 0; \\ -\pi + \tan^{-1}\left(\frac{y}{x}\right) & \text{if } x < 0 \text{ and } y < 0; \\ \pi + \tan^{-1}\left(\frac{y}{x}\right) & \text{if } x < 0 \text{ and } y > 0. \end{cases}$$

We see in Fig 6.1 that $|z|$ can be interpreted as a measure of the distance between the point z and the origin O. The formula for $|z|$ in Definition 2 can be seen to relate to Pythagora's Theorem. We can extend this a little further by considering two points in the complex plane $z_1 = x_1 + y_1 i$ and $z_2 = x_2 + y_2 i$. We then have

$$z_1 - z_2 = (x_1 - x_2) + (y_1 - y_2)i,$$
$$|z_1 - z_2| = \sqrt{((x_1 - x_2)^2 + (y_1 - y_2)^2)}.$$

The last expression gives the Pythagorean distance between the points (x_1, y_1) and (x_2, y_2) in Cartesian coordinates. We can therefore interpret $|z_1 - z_2|$ as a measure of the distance between z_1 and z_2 in the complex plane.

◉ Example 4

Describe geometrically the set of complex numbers z satisfying the equation $|z - (1 + i)| = 3$.

$|z - (1 + i)|$ measures the distance between z and $1 + i$, so we are considering the set of points in the complex plane whose distance from $1 + i$ is equal to 3. This is the circle with centre $1 + i$ and radius 3.

Example 5

Describe geometrically the set of complex numbers z satisfying $1 < |z| \leq 3$.

$|z|$ measures the distance from the origin, and so we have the set of numbers whose distance from the origin lies between 1 and 3, i.e. the region contained within the two circles, centre the origin and radii 1 and 3. The inequality symbols indicate that the outer circle is part of the set but the inner circle is not.

Example 6

Describe geometrically the set of complex numbers z satisfying $-\pi/4 < \arg(z) < \pi/4$.

Referring to Fig 6.1 tells us that we have the set of points for which the angle θ lies between $\pm\pi/4$, i.e. the infinite wedge-shaped region in the right-hand half-plane contained between the lines $y = x$ and $y = -x$. The lines themselves are excluded.

We now turn our attention to addition. The Hamilton notation for complex numbers, as well as being the same as for Cartesian coordinates, is also that used for two-dimensional vectors, and the Hamiltonian definition of addition is precisely that of vector addition. We would therefore expect a diagram for complex addition to appear the same as the parallelogram rule for vector addition. Subtraction is dealt with similarly, and we can see both in Fig 6.2. The diagram also illustrates the

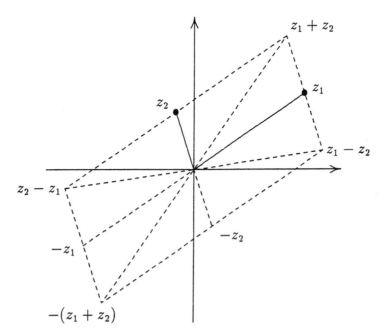

Fig 6.2 Addition and subtraction in the complex plane.

triangle inequality (Proposition 5). If we consider the triangle whose vertices are the origin, the point z_1 and the point $z_1 + z_2$, we can see that, from the properties of parallelograms, the distance from z_1 to $z_1 + z_2$ is equal to that from the origin to z_2. Using the fact that the sum of two sides of a triangle always exceeds the third side confirms that $|z_1 + z_2| \le |z_1| + |z_2|$.

TUTORIAL PROBLEM 5

Plot some complex numbers on graph paper and verify that sums and differences do follow the parallelogram rules illustrated in Fig 6.2.

If we look again at Fig 6.1, particularly at the interpretation of the modulus, and also consider the difficulties about giving a precise definition for the angle in terms of x and y, we can perhaps see that, in fact, polar coordinates will be a useful device in the complex plane, and we shall consider this next.

6.4 Polar Representation

The equations relating Cartesian and polar coordinates are

$$x = r\cos\theta \quad \text{and} \quad y = r\sin\theta.$$

We can therefore write z in polar form as

$$z = x + yi = r(\cos\theta + i\sin\theta).$$

Notice that we write $i\sin\theta$ instead of $\sin\theta i$ so as to avoid possible confusion of the last expression with $\sin(\theta i)$. Recalling the meanings of r and θ in polar coordinates we see immediately that

$$r = |z| \quad \text{and} \quad \theta = \arg(z).$$

Polar coordinates are better for considering multiplication. If we write two complex numbers z_1 and z_2 in polar form as

$$z_1 = r_1(\cos\theta_1 + i\sin\theta_1), \quad z_2 = r_2(\cos\theta_2 + i\sin\theta_2)$$

and multiply them together we obtain

$$\begin{aligned} z_1 z_2 &= r_1 r_2 (\cos\theta_1 + i\sin\theta_1)(\cos\theta_2 + i\sin\theta_2) \\ &= r_1 r_2 ((\cos\theta_1 \cos\theta_2 - \sin\theta_1 \sin\theta_2) + i(\cos\theta_1 \sin\theta_2 + \sin\theta_1 \cos\theta_2)) \\ &= r_1 r_2 (\cos(\theta_1 + \theta_2) + i\sin(\theta_1 + \theta_2)), \end{aligned}$$

using trigonometric addition formulae. This tells us that $|z_1 z_2| = r_1 r_2 = |z_1||z_2|$, which you may have verified in Cartesian form in Exercise 2 of §6.2. The formula also suggests that $\arg(z_1 z_2) = \arg(z_1) + \arg(z_2)$. This is not always true, because if θ_1 and θ_2 are both between $-\pi$ and π, it does not follow that $\theta_1 + \theta_2$ is within these limits. In fact, if it is outside then we find that $\arg(z_1 z_2)$ and $\arg(z_1) + \arg(z_2)$ differ by some multiple of 2π, and this also happens when we add more than two arguments in this way. Example 7 illustrates this.

● *Example 7*

Let $z_1 = -1 + i\sqrt{3}$ and $z_2 = -1 + i$. Express z_1 and z_2 in polar form, and investigate the relation between $\arg(z_1 z_2)$ and $\arg(z_1) + \arg(z_2)$.

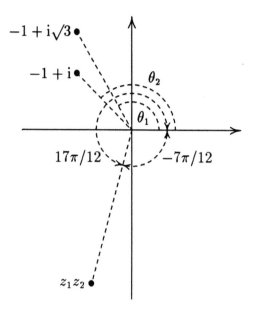

Fig 6.3 Diagram for Example 7.

From the definition of the modulus we can calculate that $|z_1| = 2$ and $|z_2| = \sqrt{2}$. From the diagram we can find the angles, so $\theta_1 = 120° = 2\pi/3$ and $\theta_2 = 135° = 3\pi/4$. This gives all the information we need, and so

$$z_1 = 2\left(\cos\frac{2\pi}{3} + i\sin\frac{2\pi}{3}\right), \qquad z_2 = \sqrt{2}\left(\cos\frac{3\pi}{4} + i\sin\frac{3\pi}{4}\right).$$

If we now use the polar product formula we obtain

$$z_1 z_2 = 2\sqrt{2}\left(\cos\frac{17\pi}{12} + i\sin\frac{17\pi}{12}\right).$$

However, $17\pi/12$ is not between $-\pi$ and π. We need to find the appropriate value for the argument and this is best done from the diagram. We have indicated the position of $z_1 z_2$ on the diagram, and we can see that the appropriate value for the angle to lie between $-\pi$ and π is $-7\pi/12$, so that $\arg(z_1 z_2) = (\arg(z_1) + \arg(z_2)) - 2\pi$ in this case.

TUTORIAL PROBLEM 6

Explore some more examples like Example 7. Try to find some general relationships between $\arg(z_1 z_2)$ and $\arg(z_1) + \arg(z_2)$, depending on which quadrants the numbers are in.

⊛ Example 8

Express the conjugate of a non-zero complex number in polar form.

From Fig 6.1 and also from the formulae in Tutorial Problem 4 we can see that $\arg(z^*) = -\arg(z)$. So if $z = r(\cos\theta + i\sin\theta)$ then
$z^* = r(\cos(-\theta) + i\sin(-\theta)) = r(\cos\theta - i\sin\theta)$.

● Proposition 6—de Moivre's theorem

If $z = r(\cos\theta + i\sin\theta)$ then $z^n = r^n(\cos(n\theta) + i\sin(n\theta))$ for all $n \in \mathbb{N}$.

PROOF

We shall give a proof of this theorem by induction. The result is certainly true for $n = 1$—there is nothing to prove. Suppose that the result is true for $n - 1$. We use the polar formula for the product of two numbers in the following calculations.

$$z^n = z \times z^{n-1}$$
$$= r(\cos\theta + i\sin\theta) \times r^{n-1}(\cos(n-1)\theta + i\sin(n-1)\theta)$$
$$= r^n(\cos\theta\cos(n-1)\theta - \sin\theta\sin(n-1)\theta$$
$$+ i\cos\theta\sin(n-1)\theta + i\sin\theta\cos(n-1)\theta)$$
$$= r^n(\cos(\theta + (n-1)\theta) + i\sin(\theta + (n-1)\theta))$$
$$= r^n(\cos(n\theta) + i\sin(n\theta)).$$

Hence by induction the result is true for all $n \in \mathbb{N}$.

TUTORIAL PROBLEM 7

Extend the polar formula for the product of two numbers to the case of n numbers, i.e. prove by induction that

$$z_1 z_2 \ldots z_n = r_1 r_2 \ldots r_n(\cos(\theta_1 + \theta_2 + \ldots + \theta_n) + i\sin(\theta_1 + \theta_2 + \ldots + \theta_n)).$$

This is a generalization of de Moivre's theorem.

EXERCISES 6.4

1. Find the modulus and argument of the numbers $1 + i$, $-i$, $1 - i\sqrt{3}$, -2, $\sqrt{3} - i$.

2. Interpret geometrically the sets of points in the complex plane satisfying:
   (i) $|z - i| = |z - 1|$,
   (ii) $|z + 1| = 2$,
   (iii) $\operatorname{Re}(z) = -3$,
   (iv) $\operatorname{Im}(z) \leq 1$.

3. Let $z_1 = -1 + i\sqrt{3}$ and $z_2 = -1 + i$. Express z_1 and z_2 in polar form, and find $\arg(z_1 z_2)$.

4. Let $z = r(\cos\theta + i\sin\theta)$. Express $1/z$ in polar form.

5. Prove the result $zz^* = |z|^2$ using the polar form of complex numbers.

6. Prove that the result of de Moivre's theorem is true for negative integers, using the theorem for positive integers and Exercise 4.

6.5 Euler's Formula

In this section we shall look at a fundamental relationship between the exponential and trigonometric functions, which bears the name of the great mathematician Euler. It is a formula which is used extensively in complex calculus. To derive it we shall need to assume some results from calculus, namely the series expansions of the exponential, sine and cosine functions.

$$e^x = 1 + x + \frac{x^2}{2!} + \frac{x^3}{3!} + \frac{x^4}{4!} + \cdots,$$

$$\cos x = 1 - \frac{x^2}{2!} + \frac{x^4}{4!} - \frac{x^6}{6!} + \cdots,$$

$$\sin x = x - \frac{x^3}{3!} + \frac{x^5}{5!} - \frac{x^7}{7!} + \cdots.$$

We now replace x in the exponential expansion by $i\theta$. We then separate out the real and imaginary parts of the resulting expression. We shall assume without proof the validity of all the algebraic manipulations we carry out on these expansions.

$$e^{i\theta} = 1 + i\theta + \frac{(i\theta)^2}{2!} + \frac{(i\theta)^3}{3!} + \frac{(i\theta)^4}{4!} + \frac{(i\theta)^5}{5!} + \frac{(i\theta)^6}{6!} + \frac{(i\theta)^7}{7!} + \cdots,$$

$$= 1 + i\theta - \frac{\theta^2}{2!} - i\frac{\theta^3}{3!} + \frac{\theta^4}{4!} + i\frac{\theta^5}{5!} - \frac{\theta^6}{6!} - i\frac{\theta^7}{7!} + \cdots,$$

$$= \left(1 - \frac{\theta^2}{2!} + \frac{\theta^4}{4!} - \frac{\theta^6}{6!} + \cdots \right) + i\left(\theta - \frac{\theta^3}{3!} + \frac{\theta^5}{5!} - \frac{\theta^7}{7!} + \cdots \right).$$

The expansions on the right are those of the trigonometric functions, and so we have Euler's Formula:

$$e^{i\theta} = \cos\theta + i\sin\theta.$$

One particular case which is often quoted is obtained by putting $\theta = \pi$. Using the fact that $\sin\pi = 0$ and $\cos\pi = -1$ enables us to deduce that

$$e^{i\pi} + 1 = 0.$$

This is thought to be remarkable because it connects in one simple formula the well-known numbers π, i, e and 1, which separately arise in quite disparate branches of mathematics. This formula has been observed emblazoned on T-shirts!

Using the laws of indices for the exponential function we can now see the polar multiplication relationship and de Moivre's theorem from a much more natural perspective. The polar form of a complex number can be written using Euler's Formula as

$$z = r(\cos\theta + i\sin\theta) = re^{i\theta}.$$

This polar exponential form is used throughout complex calculus, and we shall exploit it here. Multiplying two numbers gives

$$z_1 z_2 = r_1 e^{i\theta_1} r_2 e^{i\theta_2} = r_1 r_2 e^{i(\theta_1 + \theta_2)}.$$

The laws of indices clearly extend this to the case of n numbers, and we can derive de Moivre's theorem as follows

$$(r(\cos\theta + i\sin\theta))^n = r^n (e^{i\theta})^n = r^n e^{i(n\theta)} = r^n (\cos(n\theta) + i\sin(n\theta)).$$

Example 9

Use Euler's Formula to obtain the trigonometric identities for $\cos(A - B)$ and $\sin(A - B)$.

$$
\begin{aligned}
\cos(A - B) + i\sin(A - B) &= e^{i(A-B)} = e^{iA} e^{i(-B)} \\
&= (\cos A + i\sin A)(\cos(-B) + i\sin(-B)) \\
&= (\cos A + i\sin A)(\cos B - i\sin B) \\
&= (\cos A \cos B + \sin A \sin B) \\
&\quad + i(\sin A \cos B - \cos A \sin B).
\end{aligned}
$$

Equating real and imaginary parts then gives

$$
\begin{aligned}
\cos(A - B) &= \cos A \cos B + \sin A \sin B \\
\sin(A - B) &= \sin A \cos B - \cos A \sin B.
\end{aligned}
$$

In calculus you will meet the hyperbolic functions, defined in terms of exponentials, and Euler's Formula gives a close connection between these and the trigonometric functions. As functions of a real-variable, trigonometric and hyperbolic functions behave quite differently. The next example shows one aspect of this. As a function of a real variable, e^x is not periodic in the sense of the trigonometric functions, whereas as a function of a complex variable it is.

Example 10

Show that the exponential function is periodic with period $2\pi i$.

Using the laws of indices gives, for any complex number z,

$$e^{z + 2\pi i} = e^z e^{2\pi i} = e^z (\cos 2\pi + i\sin 2\pi) = e^z.$$

This shows that the exponential function is periodic, but it only demonstrates that $2\pi i$ is an integer multiple of the period.

Now if $e^{z+\alpha} = e^z$ then $e^\alpha = 1$. In the next tutorial problem you are asked to deduce that $\alpha = 2n\pi i$ for some $n \in \mathbb{Z}$. This shows that the period is $2\pi i$.

TUTORIAL PROBLEM 8

To show that if $e^\alpha = 1$ then $\alpha = 2n\pi i$. Write α in Cartesian form and deduce from $|e^\alpha| = 1$ that the real part of α is zero. Then use properties of trigonometric functions to deduce that the imaginary part of α is an integer multiple of 2π.

EXERCISES 6.5

1. Show that $e^{-i\theta}$ is the conjugate of $e^{i\theta}$.

2. Use Euler's Formula to obtain the trigonometric identities for $\cos(A + B)$ and $\sin(A + B)$.

3. Show that $|e^{i\theta}| = 1$ for all real numbers θ. (This shows that all such numbers lie on the unit circle centred at the origin in the complex plane.) Plot the following numbers on the unit circle,

$$e^{i\pi}, e^{i(-\pi)}, e^{i\pi/2}, e^{i\pi/3}, e^{i(-2\pi/3)}, e^{i(-\pi/2)}.$$

4. Use Euler's Formula to establish the following results,

$$\cos\theta = \frac{e^{i\theta} + e^{-i\theta}}{2}, \qquad \sin\theta = \frac{e^{i\theta} - e^{-i\theta}}{2i}.$$

6.6 The Roots of Unity

Much of our discussion about number systems has involved solving equations, and we continue that theme in this section. We mentioned in the introduction that Gauss showed that any polynomial equation has all its roots within the complex numbers, and here we shall explore the roots of the equation $z^n = 1$, where n is an arbitrary positive integer. In the cases $n = 2, 3$ these are called, respectively, the square roots and the cube roots of 1 (unity), and in general we refer to the solutions as the nth roots of unity.

We use the polar exponential form $z = re^{i\theta}$, and so $z^n = 1$ gives $r^n e^{in\theta} = 1$. Using the result of Exercise 3 in the previous section then tells us that $r^n = 1$, and since r is a non-negative real number we must have $r = 1$. So $z = e^{i\theta}$ for some value of θ, and again referring to Exercise 3 above we can see that all the roots of unity will lie on the unit circle in the complex plane. So we must have $e^{in\theta} = 1$, which tells us that $\cos(n\theta) = 1$ and $\sin(n\theta) = 0$. It follows that $n\theta$ must be an integer multiple of 2π, i.e. $\theta = 2k\pi$ for any $k \in \mathbb{Z}$. This gives the solutions of $z^n = 1$ as

$$z = e^{i\frac{2k\pi}{n}} \quad \text{for } k \in \mathbb{Z}.$$

This apparently gives us infinitely many solutions, but we shall see that different values of k do not necessarily give different values of z. This relates to Example 10. If we consider two possible values of k giving the same value of z then we have

$$z = e^{i\frac{2k_1\pi}{n}} = e^{i\frac{2k_2\pi}{n}}.$$

Dividing the two exponential expressions and using the laws of indices then gives

$$e^{i\frac{2(k_1 - k_2)\pi}{n}} = 1.$$

Using Euler's Formula and equating real and imaginary parts then gives

$$\cos\left(\frac{2(k_1 - k_2)\pi}{n}\right) = 1 \quad \text{and} \quad \sin\left(\frac{2(k_1 - k_2)\pi}{n}\right) = 0.$$

This happens if and only if $2(k_1 - k_2)\pi/n$ is an integer multiple of 2π, so that $k_1 - k_2$ must be an integer multiple of n. Therefore, there will only be n different values of k which give rise to distinct values of z. In particular, if we take $k = 0, 1, 2, \ldots, n - 1$ in turn this will give rise to the n distinct values of z. We can therefore list the distinct solutions of $z^n = 1$ as

$$1, e^{i\frac{2\pi}{n}}, e^{i\frac{4\pi}{n}}, e^{i\frac{6\pi}{n}}, \ldots, e^{i\frac{2(n-1)\pi}{n}}.$$

Recalling that these numbers all lie on the unit circle we can see that the arguments increase from one to the other by the addition of $2\pi/n$, so that they will be equally spaced around the circle, at the vertices of a regular n-sided polygon.

Example 11

Find the solutions of the equation $z^6 = 1$ and plot them in the complex plane. Express them all in Cartesian form.

The theory above tells us that the solutions will be

$$z = e^{i\frac{2k\pi}{6}} \quad (k = 0, 1, 2, 3, 4, 5)$$
$$= \cos\left(\frac{2k\pi}{6}\right) + i\sin\left(\frac{2k\pi}{6}\right).$$

These are plotted in Fig 6.4 (page 102) at the vertices of a regular hexagon, and expressed in Cartesian form using knowledge of trigonometric values.

TUTORIAL PROBLEM 9

If we now let $z_k = e^{i2k\pi/6}$ we can explore successive powers. The theory above tells us that the first six powers of z_1 will give all the distinct 6th roots of unity. Verify that the first six powers of z_5 also give all six roots. Plot and label them. Calculate successive powers of the other roots and plot and label them also, showing that in these cases we do not obtain all six roots. In particular, verify that $z_2^3 = 1$, $z_3^2 = 1$ and $z_4^3 = 1$.

Following this situation, a root of unity whose successive powers give all the roots is called a *primitive* root of unity. It turns out that $z_k = e^{i2k\pi/n}$ is a primitive nth root of unity if and only if the highest common factor of k and n is equal to 1. Try to prove this result.

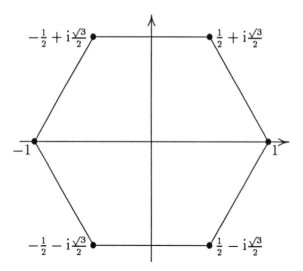

Fig 6.4 The solutions of $z^6 = 1$ in the complex plane.

Let us return to the sixth roots of unity, using the notation of Tutorial Problem 9. We can factorize $z^6 - 1$, knowing the roots, and noting that $z_0 = 1$, as

$$z^6 - 1 = (z - 1)(z - z_1)(z - z_2)(z - z_3)(z - z_4)(z - z_5).$$

We can use this to find various relationships. For example, by considering the constant term we can see that the product of the roots will be equal to -1. Another interesting result is obtained by considering the sum of the roots. This would correspond to the coefficient of z^5 on the right-hand side if we were to multiply out the brackets. This coefficient must be zero from the left-hand side, so that the sum of the roots is zero. This in fact generalizes. The algebraic reasoning is exactly the same, and so the sum of the nth roots of unity is zero. In the case when n is even it is easy to see this from the complex plane. From Fig 6.4 we can imagine that if instead of a hexagon we had a regular polygon with an even number of vertices, then every vertex would correspond to one immediately opposite, joined through the origin. For each root its negative is therefore also a root, so that they clearly all sum to zero. For the case of an odd value of n this reasoning does not work, but the result is nevertheless true, as the algebra has demonstrated. We can certainly see that the sum of the roots will be real, using Proposition 4. For any root of unity its conjugate will also be a root of unity, and when we add them the imaginary parts will cancel. If, for instance, we draw a regular pentagon in the unit circle with one vertex at 1 on the real axis we can see that it is symmetrical about the real axis, and we can use the reflective symmetry to pair off each non-real vertex with its conjugate.

Example 12

Prove that

$$1 + \cos\left(\frac{2\pi}{5}\right) + \cos\left(\frac{4\pi}{5}\right) + \cos\left(\frac{6\pi}{5}\right) + \cos\left(\frac{8\pi}{5}\right) = 0.$$

The sum of the 5th roots of unity is zero. Therefore, the sum of the real parts is also zero. The left-hand side of the equation is the sum of the real parts of the 5th roots of unity.

Finally, in this section we shall investigate the more general equation $z^n = \alpha$, where α is some complex number different from 1. If we write α in polar form as $\alpha = r e^{i\phi}$ then there is a unique positive real nth root of the positive real number r and if we write $z_0 = \sqrt[n]{r} e^{i\phi/n}$ then the laws of indices tell us that $z_0^n = \alpha$, so that we have found one solution. If we have amother solution z_1 then $(z_1/z_0)^n = 1$, so that z_1/z_0 is an nth root of unity. So all the solutions of $z^n = \alpha$ will be of the form

$$z = z_0 e^{i \frac{2k\pi}{n}} \quad (k = 0, 1, 2, \ldots, n - 1).$$

Example 13

Find the fourth roots of i. In polar exponential form we have $i = e^{i\frac{\pi}{2}}$. So one root will be $e^{i\frac{\pi}{8}}$, and therefore, using the laws of indices, the set of four will be

$$e^{i\frac{\pi}{8}}, \; e^{i\frac{\pi}{8}} \times e^{i\frac{2\pi}{4}}, \; e^{i\frac{\pi}{8}} \times e^{i\frac{4\pi}{4}}, \; e^{i\frac{\pi}{8}} \times e^{i\frac{6\pi}{4}}.$$

We can simplfy these as follows,

$$e^{i\frac{\pi}{8}}, ie^{i\frac{\pi}{8}}, -e^{i\frac{\pi}{8}}, -ie^{i\frac{\pi}{8}}.$$

They lie at the vertices of a square as in Fig 6.5.

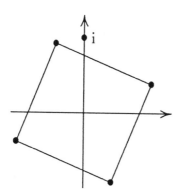

Fig 6.5 The fourth roots of i in the complex plane.

EXERCISES 6.6

1. Find the fourth roots of unity. Express them in polar exponential and Cartesian form, and plot them in the complex plane.

2. Find the cube roots of 8i. Express them in polar exponential and Cartesian form, and plot them in the complex plane.

3. Find the cube roots of unity. Let w_1, w_2 denote the two non-real roots. Show that

$$w_1^* = w_2, \; w_2^* = w_1, \; w_1^2 = w_2, \; w_2^2 = w_1.$$

Interpret these relationships in the complex plane.

4. Show that if $|z| = 1$ and $\mathrm{Re}(z) = -\frac{1}{2}$ then z is a cube root of unity.

5. Use the polar exponential form of the sixth roots of unity and the results of Exercise 4 of §6.5 to factorize $z^6 - 1$ into a product of real linear and quadratic factors.

6. Find the primitive 8th roots of unity.

Summary

The theory of complex numbers and the calculus of complex functions are among the greatest achievements of post-Renaissance mathematics. In this chapter we have introduced the basic elements of complex numbers.

In many ways the most important thing in this chapter is Euler's Formula. The geometrical representation of complex numbers, the use of polar coordinates and the complex exponential form are vital ingredients without which complex function theory and its applications would be impossible. We have explored just one instance of how powerful the polar exponential form can be in the section on the roots of unity.

In coordinate geometry, Cartesian coordinates are likely to be much more familiar than polar coordinates, but in the context of complex numbers the Cartesian form has considerable limitations. It is therefore the polar aspect of complex numbers which merits most study, and readers should aim to become fluent in this form of representation.

EXERCISES ON CHAPTER 6

1. Verify from Hamilton's definition that complex numbers obey the distributive axiom. Give a careful analysis of which algebraic rules for $\mathbb{R}$ you use at each stage of your calculations.

2. Work through the proof of Proposition 1 using the traditional notation in place of Hamilton's.

3. Extend the results of Proposition 3, proving by induction that

$$(z_1 + z_2 + \ldots + z_n)^* = z_1^* + z_2^* + \ldots + z_n^*,$$

$$(z_1 \times z_2 \times \ldots \times z_n)^* = z_1^* \times z_2^* \times \ldots \times z_n^*.$$

Deduce that

$$\left(\frac{1}{z_1 \times z_2 \times \ldots \times z_n} \right)^* = \frac{1}{z_1^*} \times \frac{1}{z_2^*} \times \ldots \times \frac{1}{z_n^*}.$$

Write down the results for the special case $z_1 = z_2 = \ldots = z_n = z$.

4. Solve the quadratic equation $iz^2 - (2 + 3i)z + (3 + i) = 0$, expressing the solutions in the form $a + bi$ where a and b are real.

5. Use de Moivre's theorem to show that

$$\cos 3A = \cos^3 A - \cos A \sin^2 A \quad \text{and} \quad \sin 3A = \cos^2 A \sin A - \sin^3 A.$$

Find a formula for $\cos 3A$ in terms of $\cos A$ only and for $\sin 3A$ in terms of $\sin A$ only.

6. Let $z = r(\cos \theta + i \sin \theta)$ denote a complex number expressed in polar form. Prove that $e^{-r} \le |e^z| \le e^r$.

7. Expressing z in polar form, find all complex numbers satisfying $|e^z| = |e^{iz}|$.

8. Find the fourth roots of $2 - 2i$. Express them in polar exponential and Cartesian form, and plot them in the complex plane.

9. Prove that

$$\sin\left(\frac{2\pi}{7}\right) + \sin\left(\frac{4\pi}{7}\right) + \sin\left(\frac{6\pi}{7}\right) + \sin\left(\frac{8\pi}{7}\right) + \sin\left(\frac{10\pi}{7}\right) + \sin\left(\frac{12\pi}{7}\right) = 0.$$

10. Find all the solutions of the equation $(z - 1)^4 = (z + 1)^4$. Simplify the resulting trigonometric expressions for z and hence show that the solutions are all purely imaginary.

11. Find all the solutions of the equation

$$z^8 - 2z^4 + 4 = 0,$$

expressing them in polar exponential form.

7 • Sequences

A sequence of numbers is a set of numbers arranged in some particular order, as when we count 1, 2, 3, 4, 5, ... Arrangements of this kind, whether of numbers, or days of the week, or students' position in an examination, are described in English using ordinal numbers, i.e. 'first', 'second' etc, and in thinking of a numerical sequence we are simply labelling the numbers in this ordinal fashion. Such activities occur very early in our mathematical development; indeed they begin with counting and number rhymes and games. Learning to count, and learning to generate patterns like 2, 4, 6, 8, 10, . . . involves the understanding of rules for moving from one place in the sequence to the next, and we considered this in the context of mathematical induction in Chapter 2. One of the activities which many pupils undertake when they are learning a multiplication algorithm is that of doubling. Here, the sequence of numbers 2, 4, 8, 16, 32, 64, . . . is generated by an inductive process whereby to obtain any number from its predecessor we multiply by 2. This has the potential of continuing without ever stopping, generating an *infinite sequence* of numbers.

One of the most important occurrences of sequences is in connection with approximations, and we encounter this with decimals. For example we know that it is impossible to express the number π as a fraction, and so if we wish to perform arithmetic using π we can only do so by substituting for π a number with which we can actually calculate. If the calculations involve decimals we may use 3.14, or maybe 3.142, depending upon the degree of accuracy required. The decimal expansion continues without stopping, and we occasionally see in the media that the latest supercomputer has calculated π to a few million more decimal places than before. The first few places are given by 3.14159265, and we would use this knowledge to generate a sequence of approximations, 3.14, 3.142, 3.1416, 3.14159, 3.141593, 3.1415927, . . . , where we hope that each number is a better approximation to π than its predecessor, in the sense that the numbers are getting closer to π as we progress along the sequence. Investigating the properties of such sequences of approximations occupied many mathematicians, among the foremost of whom were Cauchy, Euler and Weierstrass. In §5.2 we gave a very brief historical survey, mentioning the development of a theory of limits. In this context the analysis of infinite sequences was especially important, and part of this was the formulation of a mathematically adequate definition of the limit of an infinite sequence. We shall consider this together with some of the important properties of limits of sequences in this chapter.

7.1 Defining an Infinite Sequence

If we consider two of the patterns in the introduction, the even numbers and those obtained by doubling, we can express both of them through an algebraic formula. In the first case the formula $2n$ would give the nth number in the sequence, and in the second case this could be expressed by 2^n. The variable n is of course a natural number, and $n = 1, 2, \ldots$ corresponds to the first, second ... number in the sequence. So in this case any member of the sequence is expressed as a function of the variable n, the function being given by an algebraic formula, the familar way we use of expressing functional relationships. In fact, an abstract definition of a sequence is little more than that.

● *Definition I*

An infinite sequence is a function with domain $\mathbb{N}$. ●

Like many definitions in mathematics, this is very concise. It can be thought of as an abstract way of describing a sequence which encapsulates the ideas from the discussion here. It is a way of placing a set of numbers in a specified order determined by the ordering in $\mathbb{N}$ through the functional relationship. The definition says nothing about the values of the function. If they were real numbers we would have a real sequence, but they could equally well be vectors, or transformations, or some other mathematical objects. We might well want to investigate a sequence of rotations in the plane for example.

In dealing with sequences we traditionally do not use the normal functional notation. We use a suffix notation instead, and this would have been used historically before the concept of a function was seen to be as universal as we consider it now. We discussed this briefly in §5.2. So instead of writing $f(n)$ to denote the nth number in a sequence, we use a_n, b_n, or some such literal notation. We would therefore write $a_n = 2n$ to express the fact that we were considering an infinite sequence where the nth number was given by the formula $2n$. If we want to refer to the sequence as a single mathematical entity we may choose to abbreviate it as a single letter, like A, or sometimes we use a notation involving parentheses (a_n). Such notation seems rather pedantic, but would be used in situations where it is important to distinguish between the sequence as a whole and the nth member of that sequence. Often we find the notation a_n used ambiguously to denote either the sequence or its nth member, just as $f(x)$ is used for both the function and its value. Sometimes the notation a_0 is used for the first member of a sequence, rather than a_1. This may be for algebraic reasons, but is often used to indicate that a_0 is an initial guess made before a sequence of systematic approximations is generated. We shall follow this practice where appropriate. Finally, we have a piece of terminology. The numbers in a sequence are commonly referred to as the *terms* of the sequence, so that we would talk about 'the sequence whose nth term is given by $2n$'.

We mentioned mathematical induction, and both the examples above could be defined this way as follows.

For the even numbers we have $a_1 = 2$ and $a_n = a_{n-1} + 2$ for all $n \in \mathbb{N}$.

For the doubling sequence we have $a_1 = 2$ and $a_n = 2 \times a_{n-1}$ for all $n \in \mathbb{N}$.

TUTORIAL PROBLEM I

> Prove by induction that these recursive definitions do give sequences with the corresponding algebraic formulae.

Another example of a sequence defined by induction is the well-known Fibonacci sequence. This is defined inductively by

$$a_1 = 1, \quad a_2 = 1, \quad a_n = a_{n-1} + a_{n-2} \text{ for } n \geq 3.$$

This says that each term of the sequence is the sum of the previous two, where it starts with two 1s. The first few terms are 1, 1, 2, 3, 5, 8, 13, 21, 34, 55.

TUTORIAL PROBLEM 2

> Find some interesting properties of the Fibonacci sequence. Many books on number theory discuss this sequence and its applications.

7.2 Solving Equations

A common context for sequences of approximations is that of solving equations that do not have rational arithmetic solutions. We begin with an example involving $\sqrt{2}$, i.e. solving the equation $x^2 = 2$.

Example I

Find a sequence of rational numbers which gives an approximation to $\sqrt{2}$.

There are many ways of doing this, and here we choose just one. We can easily find a number whose square is quite close to 2, for example $(3/2)^2 = 2\frac{1}{4}$. So let $a_1 = 3/2$ denote the first approximation. Because $a_1^2 > 2$ we shall have $(2/a_1)^2 < 2$. So a_1 is a bit too big and $2/a_1$ is a bit too small. The average of these two numbers should therefore give us a better approximation, and so we let

$$a_2 = \frac{1}{2}\left(a_1 + \frac{2}{a_1}\right).$$

Doing the arithmetic gives $a_2 = 17/12$, a number whose square is closer to 2 than that of 3/2. We can continue this process, and use a recursive definition, letting $a_1 = 3/2$ and

$$a_n = \frac{1}{2}\left(a_{n-1} + \frac{2}{a_{n-1}}\right).$$

If we begin this procedure with $a_1 = 3/2$ we shall obtain the following table of results. The formula will always give rise to a fraction, and we have listed these and also a decimal approximation to each fraction. We have also shown the square of each approximation, again giving a decimal approximation, to give an idea of how quickly the errors decrease.

n	x_n	decimal	x_n^2	decimal
1	3/2	1.5	9/4	2.25
2	17/12	1.4166667	289/144	2.0069444
3	577/408	1.4142157	332929/166464	2.0000060

In this case we need only three steps to obtain a very good approximation.

Because such a procedure uses the same steps repeatedly, it is referred to as an *iterative* process, and it can easily be programmed. The following short segment of Pascal will produce a table of decimal approximations (where the variable x is of type real). Expert programmers might like to expand it to produce the fractional approximations also.

```
x:=2; n:=1;
repeat
writeln(n,x,x*x);   (properly formatted)
x:=(x+2/x)/2;
until abs(x*x-2)<1E-6;   (or some other stopping condition).
```

TUTORIAL PROBLEM 3

> Perform the fourth iteration in Example 1. The exact arithmetic may be beyond your calculator, so try to devise a way of working out the integer arithmetic exactly but economically, using your calculator where it will give exact partial answers.

If we replace the approximations in the recursive definition

$$a_n = \frac{1}{2}\left(a_{n-1} + \frac{2}{a_{n-1}}\right)$$

by the variable x we obtain the equation

$$x = \frac{1}{2}\left(x + \frac{2}{x}\right).$$

When this equation is simplified it yields $x^2 = 2$, whose positive solution is the number which we have been approximating. This procedure can be generalized, so that to solve an equation we try to rearrange it in the form $x = F(x)$, and then generate an iterative sequence defined by $a_n = F(a_{n-1})$, with a first approximation which we may guess from a graph. We shall explore this procedure in some detail with another example.

Example 2

Use iteration to find approximate solutions for $x^3 - 5x + 3 = 0$.

The equation can be rearranged in the form $x = F(x)$ in many ways, and we explore four of them here, namely

$$x = \frac{x^3 + 3}{5}, \qquad x = x^3 - 4x + 3, \qquad x = \frac{5x - 3}{x^2}, \qquad x = \sqrt{\frac{5x - 3}{x}}.$$

If we look at a graph of the original cubic (Fig 7.1) we can see that it has three roots, near to 1.8, 0.6 and -2.5. Using each of the iterations associated with the four rearrangements gives a variety of results, some of which are in the table below. Graphs of the various rearrangements are also given, in Fig 7.2. In each case the graph of $y = x$ has been drawn (the axis scales are unequal). The intersection of $y = x$ with $y = F(x)$ gives $x = F(x)$, which is the equation giving rise to the iteration. From the graphs it can be seen that all these intersections correspond to roots of the original cubic equation. All the graphs in this example were initally explored on the computer package Graphical Calculus which can solve equations iteratively using sequences derived from $x = F(x)$.

In the next section we shall begin to analyse the approximation errors involved. The table below shows the variations in the number of steps needed to achieve a prescribed degree of accuracy.

$F(x)$	Initial value	Root	No. of steps
$\dfrac{x^3 + 3}{5}$	$a_0 = 1$	0.65662043	16
$\dfrac{x^3 + 3}{5}$	$a_0 = -2.5$	out of range	8
$\dfrac{x^3 + 3}{5}$	$a_0 = -2.4$	0.65662043	17
$x^3 - 4x + 3$	$a_0 = 1$	out of range	6
$x^3 - 4x + 3$	$a_0 = -2.5$	out of range	4
$\dfrac{5x - 3}{x^2}$	$a_0 = 1$	1.83424318	31
$\dfrac{5x - 3}{x^2}$	$a_0 = 0.4$	out of range	14
$\sqrt{\dfrac{5x - 3}{x}}$	$a_0 = 1$	1.83424318	15

Note that we have used a_0 to denote the initial guess, as discussed above. The entries 'out of range' describe a situation where the sequence generated becomes very large in magnitude, and is soon beyond the numbers a computer can deal with. A full appreciation of this example can only be obtained by exploring it on a computer or a graphics calculator.

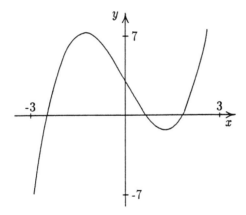

Fig 7.1 Graph of $y = x^3 - 5x + 3$.

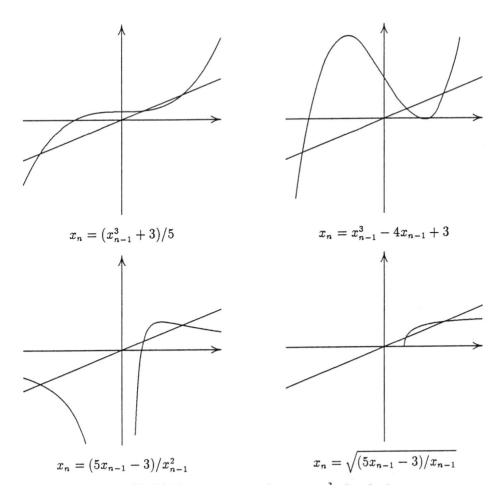

$$x_n = (x_{n-1}^3 + 3)/5$$

$$x_n = x_{n-1}^3 - 4x_{n-1} + 3$$

$$x_n = (5x_{n-1} - 3)/x_{n-1}^2$$

$$x_n = \sqrt{(5x_{n-1} - 3)/x_{n-1}}$$

Fig 7.2 Various iterations for solving $x^3 - 5x + 3 = 0$.

EXERCISES 7.2

1. Use the method of Example 1 to find a sequence of rational approximations to $\sqrt{5}$. Write a computer program to generate the sequence.

2. Rearrange the following equations in the form $x = F(x)$ in as many ways as you can think of.

   (i) $x^3 - 6x - 4 = 0$,   (ii) $x^2 = \sin x$,   (iii) $e^x = 3x^2 - 2$.

   Use a graphics calculator or a computer package to draw graphs of $y = F(x)$ and $y = x$. Investigate the behaviour of the associated sequences given by $a_n = F(a_{n-1})$, choosing various initial values. Draw up tables of results as in Example 2.

3. A chord is drawn in a circle in such a way as to cut off a segment whose area is equal to one third of the area of the whole circle. Find, to as many decimal places as you can, the distance from the centre of the circle to the centre of the chord.

7.3 Limits of Sequences

In §5.2 we saw that the requirement for a mathematical analysis of limits became a necessity during the 19th century. We shall analyse one of the illustrations used in Example 2 above to indicate some of the features which contribute to an abstract definition of the limit of a sequence. If iterations are explored using a computer package it becomes clear that there are many situations where the successive approximations become closer and closer to some number which a graph will have told us corresponds to a root of the equation being studied. We refer to this number as the *limit* of the sequence. It is a value which is approximated by considering a_n as n increases. It is conventional to say that we are considering the limit *as n tends to infinity*. This language, historically derived, has its dangers, for it is tempting to think that there is a number 'infinity' towards which n tends. We shall mostly try to avoid this terminology and simply talk about the limit of a sequence.

Consider the iteration which occurred in Example 2 given by

$$a_n = \frac{a_{n-1}^3 + 3}{5}.$$

We saw that with a suitable choice of a_0 the sequence approximated to a limit in the region of 0.65. Let us choose a_0 between 0.6 and 0.7, and investigate bounds for a_1. Now

$$a_1 = \frac{a_0^3 + 3}{5} \quad \text{so that} \quad \frac{0.6^3 + 3}{5} < a_1 < \frac{0.7^3 + 3}{5},$$

giving $0.6 < 0.6432 < a_1 < 0.6686 < 0.7$.

In fact the same algebra shows that if $0.6 < a_{n-1} < 0.7$ then $0.6 < a_n < 0.7$ and hence by induction that all members of the sequence lie between 0.6 and 0.7 if the first term is in this interval. This information is numerically fairly crude, but we can refine it as follows. We first note that the limit l of the sequence also lies between 0.6 and 0.7, and is a root of the equation, therefore satisfying $l = (l^3 + 3)/5$. We now have

$$|a_n - l| = \left| \frac{a_{n-1}^3 + 3}{5} - l \right| = \left| \frac{a_{n-1}^3 + 3}{5} - \frac{l^3 + 3}{5} \right|$$

$$= \left| \frac{a_{n-1}^3 - l^3}{5} \right| = |a_{n-1} - l| \left| \frac{a_{n-1}^2 + a_{n-1}l + l^2}{5} \right|$$

$$< |a_{n-1} - l| \times \frac{0.7^2 + 0.7^2 + 0.7^2}{5}$$

$$= 0.294|a_{n-1} - l|.$$

So applying this inequality repeatedly gives

$$|a_n - l| < 0.294|a_{n-1} - l| < (0.294)^2|a_{n-2} - l|$$
$$< (0.294)^3|a_{n-3} - l| < (0.294)^n|a_0 - l| < (0.294)^n \times 0.1.$$

Now suppose we want to find how many terms of the sequence we need to generate to achieve an accuracy of 10^{-7} in the approximation. The last set of inequalities tells us that this will be attained provided $(0.294)^n \times 0.1 < 10^{-7}$, i.e. $(0.294)^n < 10^{-6}$. To solve this inequality we need to apply the logarithmic function to both sides (see Example 11 and Exercise 1 of §4.5). This gives

$$n \ln(0.294) < -6 \ln(10), \quad \text{so} \quad n > \frac{-6 \ln(10)}{\ln(0.294)} \approx 11.3.$$

(Notice that in the last inequality we have divided by the negative number $\ln(0.294)$ which is why the inequality sign has become reversed.) So we need $n \geq 12$. This means that we can be certain of the desired accuracy from the 12th term onwards. In fact, if the iteration is performed 12 times a somewhat better accuracy will be obtained. This is because we have been using the fairly crude relationship $|a_n - l| < 0.294|a_{n-1} - l|$. Nevertheless, from this inequality we have been able to obtain some valuable information. It is worth noting that this analysis did not require us to know the exact value of l. This will be the case for many sequences, where we shall need to determine whether or not they have a limit without necessarily being able to find or guess in advance what that limiting number will be.

TUTORIAL PROBLEM 4

How far along the sequence above do we need to go in order to achieve an accuracy of 10^{-50}?

The two measures of accuracy, 10^{-7} and 10^{-50}, could be replaced by an arbitrary measure, as small as we like. If that measure were denoted by the Greek letter ϵ (read as 'epsilon') we would have to solve the inequality $(0.294)^n \times 0.1 < \epsilon$, giving

$$n > \frac{\ln(\epsilon) - \ln(0.1)}{\ln(0.294)}.$$

So if we let N denote the smallest integer which is greater than the number on the right-hand side then we shall have accuracy better than ϵ for all terms of the sequence for which $n \geq N$.

The discussion here concerning measures of accuracy covers precisely those features which comprise the abstract definition of the limit of a sequence.

● Definition 2

A sequence (a_n) has limit l if, for an arbitrary positive measure of accuracy ϵ, there is a position N in the sequence from which all the terms of the sequence satisfy $|a_n - l| < \epsilon$, i.e. this inequality is satisfied for all values of n which exceed N. We can formulate this as a logical statement involving quantifiers as follows:

$$\forall \epsilon > 0, \exists N \in \mathbb{N}, \forall n \geq N, |a_n - l| < \epsilon. \qquad ●$$

A sequence which has a limit is said to *converge*. A sequence which does not have a limit is said to *diverge*. We use two notations, either $\lim(a_n) = l$ read as 'the limit of the sequence (a_n) is l', or $a_n \to l$ read as 'a_n tends to l'.

This symbolic statement appears somewhat forbidding, and has to be understood not only through the verbal equivalent in the definition, but also through the previous discussion and through the work in the rest of this chapter. It is abstract and logically complex, as is to be expected from something which took a couple of hundred years to evolve into this formulation.

Having considered the definition verbally and symbolically, and in Example 2 numerically, we shall give a graphical interpretation. In Fig 7.3 we have plotted the terms of the sequence as points (n, a_n). The limit is shown by drawing a horizontal line at height l. The figure is meant to indicate that terms a_n for which $n \geq N$ all lie within ϵ of l, so they lie between the two dashed lines. For preceding terms, some may be within the dashed lines and some not. Notice that there may be terms of the sequence actually equal to l. The use of language such as 'tends to' and 'limit'

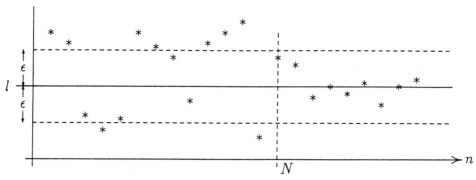

Fig 7.3 Graphical representation for $\lim(a_n) = l$.

suggests that this should perhaps not be the case. In fact, if we consider the sequence, all of whose terms are equal to 2, according to the definition the limit of this sequence is 2, because $|a_n - 2| = 0$ for all n, and so certainly $|a_n - 2| < \epsilon$ for any $\epsilon > 0$. We shall return to this point after the proof of Proposition 1.

In order to apply the definition to particular sequences we do sometimes have to try to guess the value for the limit l, and so you are encouraged to explore on a computer or calculator the successive terms of any sequence considered in order to try to guess the limit.

TUTORIAL PROBLEM 5

It is implicit in the discussion above, and in the rest of the chapter, that limits are unique, i.e. that a sequence cannot have more than one limit. Notice however that nothing in the definition states this. Construct a proof from the definition by filling in the details of the following outline.

Suppose a sequence (a_n) has two limits, l and m. We use the triangle inequality to say that

$$|l - m| = |(a_n - m) - (a_n - l)| \leq |a_n - m| + |a_n - l|.$$

From the definition each of the two terms on the right can be made as small as we like by ensuring that n is sufficiently large. This shows that the non-negative number $|l - m|$ is smaller than any positive number, and so it must be zero, proving that $l = m$.

In Example 2 we found some situations where the terms of the sequence of approximations became very large, and it seemed clear that, apart from the limitations of computer arithmetic, this would continue with no upper bound, that the terms could become arbitrarily large, and remain so. The traditional language for describing this situation is to say that the sequence tends to infinity, again with the danger that we may be led into thinking of the word 'infinity' as denoting a number. When we look at the definition of this behaviour we see that this is not the case. The symbolism used is $a_n \to \infty$, and the use of the symbol ∞ to stand for 'infinity' is even more dangerous in this regard.

● Definition 3

A sequence (a_n) is said to tend to infinity if, however large a number is specified, there is a position in the sequence from which all the terms exceed that number. Symbolically we have

$$\forall K \in \quad , \exists N \in \quad , \forall n \geq N, a_n > K. \qquad ●$$

Example 3

Prove from Definition 3 that the sequence defined by $a_n = \ln(n + 1)$ tends to infinity.

Let K denote an arbitrary real number. We want to solve the inequality $a_n > K$, in this case $\ln(n+1) > K$. This is satisfied if and only if $n + 1 > e^K$, i.e. $n > e^K - 1$. So if we let N denote the least integer greater than $e^K - 1$ we have shown that for all values of $n \geq N$ we have $a_n > K$.

Note that we have proved the existence of N by giving an expression for N in terms of K. The order of quantification allows this dependence of N upon K, as was discussed in §1.5. Note also that the definition does not require us to find the smallest such K.

TUTORIAL PROBLEM 6

> Formulate a definition of a sequence tending to $-\infty$, i.e. the terms becoming arbitrarily large and negative. Give an example and show that it satisfies your definition.

Having given definitions which, certainly in their symbolic form, seem very abstract, we need to see whether they are sensible and useful, i.e. that

(i) particular sequences which intuition tells us have limits are assigned these limits by the definition,

(ii) the definitions can be used to prove results about limits, again which intuition and examples tell us should be valid.

An example of (ii) involves the idea that if one sequence of numbers becomes close to l and another sequence of numbers becomes close to m, then adding corresponding terms should give us a sequence whose limit is $l + m$. This is one aspect of the algebra of sequences, contained in the next proposition.

● *Proposition I—The algebra of sequences*

Suppose $\lim(a_n) = l$ and $\lim(b_n) = m$. Then

(i) $\lim(a_n + b_n) = l + m$,

(ii) $\lim(k + a_n) = k + l$ for any $k \in \mathbb{R}$,

(iii) $\lim(k \times a_n) = k \times l$ for any $k \in \mathbb{R}$,

(iv) $\lim(a_n - b_n) = l - m$,

(v) $\lim(a_n b_n) = lm$,

(vi) $\lim(1/b_n) = 1/m$ provided $b_n \neq 0, m \neq 0$,

(vii) $\lim(a_n/b_n) = l/m$ provided $b_n \neq 0, m \neq 0$. ●

Before embarking on the proof we shall give an illustration of the use to which this proposition is put. We shall determine the limit of the sequence (a_n) given by

$$a_n = \frac{2n^2 + 3n - 5}{3n^2 - n + 7},$$

using the result that $\lim(1/n) = 0$. We rewrite a_n as

$$a_n = \frac{2 + 3/n - 5/n^2}{3 - 1/n + 7/n^2}.$$

Now using (v) tells us that $\lim(1/n^2) = \lim(1/n) \times \lim(1/n) = 0$. Then from (iii) we deduce that $\lim(5/n^2) = \lim(7/n^2) = \lim(3/n) = 0$. Using (iv) and (ii) then gives $\lim(2 + 3/n - 5/n^2) = 2$ and $\lim(3 - 1/n + 7/n^2) = 3$. Finally, applying (v) gives $\lim(a_n) = 2/3$.

PROOF

The proofs in this proposition are the most logically complex in this book, and so we have included some discussion alongside the formal aspect, to explain something of the underlying intuition. The proofs need analysis of statements involving quantifiers and a detailed understanding of the relationships between the variables used. We have made the point already that this area of mathematics took the most able mathematicians of the 19th century a great deal of effort to develop. Whilst drawing this to the reader's attention I make no apology for including a small amount of logically challenging work alongside the more straightforward examples and methods.

(i) The central idea here is that we know something about the size of $|a_n - l|$ and $|b_n - m|$ and from this we try to deduce something about the size of $|(a_n + b_n) - (l + m)|$. By 'size' we are referring to the idea of a measure of accuracy considered in the discussion leading up to Tutorial Problem 4. So we are trying to express what we *want* to know in terms of what we *already* know, as we do throughout mathematics. Intuitively, if we know that a_n is close to l and b_n is close to m, both to within a measure of accuracy η, we would expect $a_n + b_n$ to be within 2η of $l + m$. Suppose for example that we wanted now to ensure that $a_n + b_n$ was within 10^{-7} of $l + m$. We would then expect to have to choose $\eta = 10^{-7}/2$. In Definition 2, ϵ represents an arbitrary measure of accuracy, exemplified by the number 10^{-7} above. We have then specified η in terms of ϵ. This is the essential feature of the logical relationship between the variables ϵ and η as they appear in the formal presentation of the proof. The final thing we have to realize is that in Definition 2 the letter ϵ could be replaced by α, β, η or indeed any symbol, without altering the meaning of the statement at all. This idea was discussed briefly in §1.4. We present the formal proof as follows.

Let ϵ denote an arbitrary positive number. Let $\eta = \epsilon/2$. Since $\lim(a_n) = l$ and $\lim(b_n) = m$, from the definition there are natural numbers N_1 and N_2 for which $|a_n - l| < \eta$ for all $n \geq N_1$ and $|b_n - m| < \eta$ for all $n \geq N_2$. Let N denote the larger of the two numbers N_1 and N_2. Then both inequalities are satisfied for $n \geq N$. Using the triangle inequality we have

$$|(a_n + b_n) - (l + m)| = |(a_n - l) + (b_n - m)| \leq |a_n - l| + |b_n - m| < \eta + \eta = \epsilon.$$

(ii) The idea here is the straightforward geometrical notion that if we shift both a_n and l by an amount k then the distance between them remains unchanged.

Let ϵ denote an arbitrary positive number. From the definition, since $\lim(a_n) = l$, there is a number $N \in \mathbb{N}$ such that for all $n \geq N, |a_n - l| < \epsilon$. So we can deduce that for all $n \geq N, |(k + a_n) - (k + l)| = |a_n - l| < \epsilon$. This shows that $\lim(k + a_n) = k + l$.

(iii)   Again if we know that a_n is within the measure of accuracy η of l, we would expect ka_n to be within $k\eta$ of kl (in the case $k > 0$). So to achieve an accuracy of ϵ we would need to choose η to satisfy $k\eta = \epsilon$. This can be done, for Definition 2 specifies that its requirements can be met for *any* measure of accuracy.

Let ϵ denote an arbitrary positive number. From the definition, since $\lim(a_n) = l$, corresponding to any number $\eta > 0$ there is a number $N \in \mathbb{N}$ such that for all $n \geq N, |a_n - l| < \eta$. Now if $k \neq 0$ we let $\eta = \epsilon/|k|$. So for all $n \geq N, |ka_n - kl| < |k|\eta = \epsilon$. This shows that $\lim(ka_n) = kl$. If $k = 0$ then $ka_n = 0$ for all n and so $\lim(ka_n) = 0 = kl$.

(iv)   This follows from (i) and (iii). We first deduce from (iii), by taking $k = -1$, that $\lim(-b_n) = -m$. We then use (i) with the two sequences (a_n) and $(-b_n)$ to show that $\lim(a_n + (-b_n)) = l + (-m)$ which gives the result we want.

(v)   As before, we know that $|a_n - l|$ and $|b_n - m|$ can be made as small as we please, for sufficiently large values of n. We have to demonstrate that this is true of $|a_n b_n - lm|$. The algebraic relationship is not so obvious here. A reasonable strategy is to multiply $a_n - l$ and $b_n - m$, since we are aiming at a product. This gives

$$(a_n - l)(b_n - m) = a_n b_n + lm - a_n m - b_n l.$$

We want $a_n b_n - lm$, whereas the above expression contains $+lm$. This suggests that we rewrite the right-hand side as

$$a_n b_n - lm + lm + lm - a_n m - b_n l = (a_n b_n - lm) - m(a_n - l) - l(b_n - m).$$

Putting this information together gives

$$a_n b_n - lm = (a_n - l)(b_n - m) + m(a_n - l) + l(b_n - m).$$

We are aiming for a measure of accuracy signified by ϵ, and we could achieve this if we were able to ensure that each of the three factors on the right-hand side was at most $\epsilon/3$. If $|a_n - l|$ and $|b_n - m|$ were both less than $\sqrt{(\epsilon/3)}$ this would deal with the first factor. If $|a_n - l|$ were less than $\epsilon/|m|$ this would take care of the second factor, and likewise with the third factor if $|b_n - m|$ were less than $\epsilon/|l|$. But we can achieve any measure of accuracy we like for $|a_n - l|$ and $|b_n - m|$ through Definition 2. In this discussion we have divided by both l and m, and we shall deal only with the case where neither l nor m is zero, leaving the exceptions to Tutorial Problem 7. The formal proof begins in the same style as the previous parts, the written conventions of style being a part of the process of proof.

Let ϵ denote an arbitrary positive number. We now specify variables α and β in terms of ϵ.

So let $\alpha = \min\left(\sqrt{\dfrac{\epsilon}{3}}, \dfrac{\epsilon}{3|m|}\right)$ and $\beta = \min\left(\sqrt{\dfrac{\epsilon}{3}}, \dfrac{\epsilon}{3|l|}\right)$, where 'min' indicates the smaller of the two numbers in parentheses. Since $\lim(a_n) = l$ and $\lim(b_n) = m$, there are integers N_1 and N_2 such that for all $n \geq N_1, |a_n - l| < \alpha$ and for all $n \geq N_2, |b_n - m| < \beta$. Now let N denote the larger of N_1 and N_2. For all $n \geq N$ we then have

$$|a_n b_n - lm| = |(a_n - l)(b_n - m) + m(a_n - l) + l(b_n - m)|$$
$$\leq |a_n - l||b_n - m| + |m||a_n - l| + |l||b_n - m|$$
$$< \sqrt{\frac{\epsilon}{3}} \times \sqrt{\frac{\epsilon}{3}} + |m|\frac{\epsilon}{3|m|} + |l|\frac{\epsilon}{3|l|} = \epsilon.$$

This shows that $\lim(a_n b_n) = lm$.

(vi)   For this part of the proof we suggest that readers attempt to devise an informal strategy for dealing with $|(1/b_n) - (1/m)|$ along the lines of the earlier discussions. A formal proof is as follows.

Since $\lim(b_n) = m$ we know from the definition that if $\epsilon = |m|/2$ then there is a number $N_1 \in \mathbb{N}$ such that for all $n \geq N_1, |b_n - m| < |m|/2$. It follows that

$$|b_n| = |b_n - m + m| \geq |m| - |b_n - m| > |m|/2.$$

This implies, in particular, that for $n \geq N_1, b_n \neq 0$. Also, if η is any positive number, there is a corresponding integer N_2 such that for all $n \geq N_2, |b_n - m| < \eta$. Now let ϵ denote an arbitrary positive number, and let $\eta = \epsilon|m|^2/2$. Let N denote the larger of N_1, N_2. Then for all $n \geq N$ all the inequalities above hold and so

$$\left|\frac{1}{b_n} - \frac{1}{m}\right| = \frac{|b_n - m|}{|b_n||m|} < \frac{|b_n - m|}{\frac{1}{2}|m||m|} < \frac{\eta}{\frac{1}{2}|m|^2} = \epsilon.$$

This shows that $\lim(1/b_n) = 1/m$.

(vii)   This follows by using result (v) and then replacing the sequence (b_n) by the sequence $(1/b_n)$, using the result of (vi). ●

Note that one consequence of (iv) is that if we consider the special case where the sequences (a_n) and (b_n) are the same, we obtain the result that the sequence, all of whose terms are zero, has limit zero. Using this sequence in (ii) then tells us that the sequence, all of whose terms have the constant value k, has limit k. We referred to this point just after Definition 2 above.

TUTORIAL PROBLEM 7

Try to construct a proof of the cases in (v) which were omitted. Try to show that if $\lim(a_n) = 0$ and $\lim(b_n) = m$ then $\lim(a_n b_n) = 0$.

An important result involves the comparison of a sequence with other sequences having known limits.

● *Proposition 2—The sandwich theorem*

(i) Suppose that (a_n) and (b_n) are two sequences having the same limit l. Suppose that (x_n) is a sequence with the property that there is a number $M \in \mathbb{N}$ such that for all $n \geq M$, $a_n \leq x_n \leq b_n$ (i.e. x_n is 'sandwiched' between a_n and b_n). Then the sequence x_n also has the limit l.

(ii) If $\lim(b_n) = 0$ and if there is a number $M \in \mathbb{N}$ such that for all $n \geq M$, $|x_n| \leq |b_n|$. Then $\lim(x_n) = 0$.   ●

PROOF

This result is clear intuitively. If we have two sequences whose terms get very close to l, then a sequence whose terms are ultimately sandwiched between those of the first two must also get very close to l. We have commented that a test of a good definition is whether it gives rise to results such as this one which are intuitively obvious.

(i)   Let ϵ denote an arbitrary positive number. From Definition 2, there is a number $N_1 \in \mathbb{N}$ such that for all $n \geq N_1, |a_n - l| < \epsilon$. For the purposes of this proof we need to write $|a_n - l| < \epsilon$ in the equivalent form $-\epsilon < a_n - l < \epsilon$. Again from Definition 2 there is a number $N_2 \in \mathbb{N}$ such that for all $n \geq N_2, -\epsilon < b_n - l < \epsilon$. Now let N denote the larger of N_1, N_2. Then for all $n > N$ we have
$x_n - l = (x_n - a_n) + (a_n - l) \geq (a_n - l) > -\epsilon$, using the fact that $x_n \geq a_n$. Similarly, for all $n \geq N$ we have $x_n - l = (x_n - b_n) + (b_n - l) \leq (b_n - l) < \epsilon$, using the fact that $x_n \leq b_n$. We have now shown that for all $n \geq N, -\epsilon < x_n - l < \epsilon$, i.e. that $\lim(x_n) = l$.

(ii)   This is a special case of (i) obtained by considering $-|b_n| \leq x_n \leq |b_n|$ and $l = 0$. It is used sufficiently often to be highlighted in this way.   ●

In Proposition 2 the relationship between the sequences is required not for all n, but only for all n beyond a certain stage. An alternative way of expressing this is to say that we require the relationship to be satisfied for all n with a finite number of exceptions. Changing a finite number of terms of a sequence will not affect whether it converges, or the value of its limit.

In finding the limits of sequences whose terms are given by particular formulae we shall rely on a small number of standard sequences together with the algebraic results in Proposition 1, and the technique of Proposition 2, often using part (ii). We shall therefore move away from having to use the abstract definition all the time.

● *Proposition 3—Some standard sequences*

(i) For every $p > 0$, $\lim(n^{-p}) = 0$.

(ii) For every $x > 0$, $\lim(\sqrt[n]{x}) = 1$.

(iii) $\lim(\sqrt[n]{n}) = 1$.

(iv) For every x for which $|x| < 1$, $\lim(x^n) = 0$.

(v) For every x for which $|x| < 1$, and every $p > 0$, $\lim(n^p x^n) = 0$.   ●

PROOF

(i) Let ϵ denote an arbitrary positive number. Then $n^{-p} < \epsilon$ if and only if $n > \epsilon^{-(1/p)}$, this inequality being simply a rearrangement of the first. So let N denote the smallest integer greater than $\epsilon^{-(1/p)}$, and then for all $n \geq N$ we have shown that $|n^{-p} - 0| < \epsilon$. Note that $n^{-p} > 0$ for all n, and so $|n^{-p} - 0| = n^{-p}$. Some special cases we often use are $p = 1$, giving $\lim(1/n) = 0$, and other integer values of p, for example $p = 3$ giving $\lim(1/n^3) = 0$. Results such as this can be confirmed numerically using your calculator and trying large values of n in the formulae.

(ii) Firstly if $x = 1$ we have the constant sequence all of whose terms are equal to 1, which has limit 1.

Now suppose $x > 1$, and let $y_n = \sqrt[n]{x} - 1$, so that $y_n > 0$. Rearranging this equation gives $x = (1 + y_n)^n \geq 1 + ny_n$ using the binomial theorem and the fact that $y_n > 0$. Rearranging the inequality gives $0 < y_n \leq (x - 1)/n$. Using (i) with $p = 1$ and result (iii) of Proposition 1 with $k = x - 1$ tells us that $\lim((x - 1)/n) = 0$, and so by Proposition 2(ii) $\lim(y_n) = 0$. But $\sqrt[n]{x} = 1 + y_n$ and so $\lim(\sqrt[n]{x}) = 1$, using Proposition 1(ii).

Now if $0 < x < 1$ then $1/x > 1$ and so $\lim(\sqrt[n]{1/x}) = 1$. Hence, $\lim(1/\sqrt[n]{1/x}) = 1$ and so $\lim(\sqrt[n]{x}) = 1$ in this case also.

(iii) We know that $\sqrt[n]{n} > 1$ and so we can write $\sqrt[n]{n} = 1 + t_n$ where $t_n > 0$. Using the binomial expansion gives

$$n = (1 + t_n)^n \geq \frac{n(n-1)}{2!} t_n^2, \quad \text{and so} \quad 0 < t_n < \sqrt{\frac{2}{n-1}}.$$

Now $\lim\left(\sqrt{2/(n-1)}\right) = 0$ and so by Proposition 2(ii) $\lim(t_n) = 0$, so that $\lim(\sqrt[n]{n}) = 1$.

(iv) If $x = 0$ we obtain the constant zero sequence, and the result is trivial. If $0 < |x| < 1$, we can write $|x| = 1/(1 + c)$, where $c > 0$. So $|x|^n = 1/(1 + c)^n \leq 1/nc$, using the binomial theorem. But $\lim(1/nc) = 0$ and so $\lim(x^n) = 0$ by Proposition 2(ii).

(v) Let $a_n = n^p x^n$. Then

$$\lim\left|\frac{a_{n+1}}{a_n}\right| = \lim\left(\left|\left(\frac{n+1}{n}\right)^p \frac{x^{n+1}}{x^n}\right|\right) = \lim\left(\left(1 + \frac{1}{n}\right)^p |x|\right) = |x|.$$

We now let $A = \frac{1}{2}(1 + |x|)$ so that A is half way between $|x|$ and 1, and $0 < A < 1$. Letting $\epsilon = A - |x|$ ensures that $\epsilon > 0$, and Definition 2 guarantees that there is an integer N such that for all $n \geq N$, $|a_{n+1}/a_n| < A$. Then for all $n \geq N$

$$|a_n| = \left|\frac{a_n}{a_{n-1}} \frac{a_{n-1}}{a_{n-2}} \cdots \frac{a_{N+1}}{a_N} a_N\right| \leq A^{n-N} a_N = A^n \times (A^{-N} a_N).$$

Now $(A^{-N} a_N)$ is a constant, and $\lim(A^n) = 0$ using (iv), since $0 < A < 1$. Therefore, by Proposition 2(ii), $\lim(a_n) = 0$. ●

Example 4

Prove that if $a_n > 0$ for all n and if $\lim(a_n) = 0$ then $(1/a_n) \to \infty$.

Let K be an arbitrarily large positive number, and let $\epsilon = 1/K$. Since $\lim(a_n) = 0$, there is an integer N such that for all $n \geq N, |a_n| < \epsilon$. Using the fact that $a_n > 0$ enables us to deduce that for all $n \geq N, (1/a_n) > 1/\epsilon = K$. Looking back to Definition 3, this shows that $(1/a_n) \to \infty$.

Example 5

Discuss the behaviour of the sequences defined by $a_n = 1/n \cos(n\pi)$, and $b_n = n \cos(n\pi)$.

When investigating the behaviour of sequences it is sometimes helpful to write out explicitly the first few terms from the general formula, to gain a better understanding of the pattern the sequence follows. In this case for (a_n) the first few terms are $-1, +1/2, -1/3, +1/4, -1/5$, and this should be sufficient to appreciate what is happening, and that the sequence looks as if it has limit zero. To justify this we note that $\cos(n\pi) = \pm 1$, so that $|a_n| \leq 1/n$ (in fact we have equality). The sequence $(1/n)$ has limit zero, using Proposition 3(i) with $p = 1$, and so by Proposition 2(ii) $\lim(a_n) = 0$.

We now turn to b_n. The first few terms are $-1, +2, -3, +4, -5$, and it is clear from this that the sequence has no limit. It is also the case that the sequence does not tend to infinity, for although the terms become very large in magnitude they alternate in sign, so that the subsequence $(2, 4, 6, \ldots)$ tends to infinity whereas the subsequence $-1, -3, -5, \ldots$ tends to minus infinity. If we want a concise description of the overall behaviour we could say that it oscillates unboundedly.

The most important thing about this example is that it shows that $\lim(a_n) = 0$ does not imply that $(1/a_n)$ tends to infinity. This is a common mistake, and we should observe that in Example 4 the conclusion required the additional condition that the terms of the original sequence should be positive.

Example 6

Determine the limiting behaviour of the sequences defined by the following formulae

(i) $\dfrac{n^2 + 3n - 4}{2n^2 - 4n + 6}$, (ii) $\dfrac{n^2}{2^n}$, (iii) $\dfrac{5^n}{n^{(3/2)}}$, (iv) $\sqrt{n^2 + n} - n$.

(i) A general strategy in examples involving algebraic formulae of this kind with sums, quotients etc. is to look for the component which tends to infinity most rapidly and then divide it into top and bottom. In this case the dominant component is n^2, and so we write

$$\frac{n^2 + 3n - 4}{2n^2 - 4n + 6} = \frac{1 + 3/n - 4/n^2}{2 - 4/n + 6/n^2} \to \frac{1 + 3 \times 0 - 4 \times 0}{2 - 4 \times 0 + 6 \times 0} = \frac{1}{2}.$$

We have used the fact that negative powers of n tend to zero (Proposition 3(i)) and then the algebra of limits (Proposition 1). As we do more examples the need to refer explicitly to these propositions for justification will diminish through familiarity.

(ii)   This is a special case of Proposition 3(v) with $p = 2$ and $x = 2$. It is therefore what we have classified in Proposition 3 as a standard type of sequence, whose behaviour we shall simply quote when we need it. Notice that both numerator and denominator separately tend to infinity, and the fact that the quotient tends to zero can be expressed by saying that the denominator tends to infinity more quickly than the numerator, or 'exponentials tend to infinity faster than powers'. This is an informal but useful way of expressing the general result of Proposition 3(v).

(iii)   The same reasoning as in (ii) tells us that $(n^{(3/2)}/5^n) \to 0$, and now the fact that the terms are all positive enables us to use Example 4 to say that $(5^n/n^{(3/2)}) \to \infty$. Again exponentials tend to infinity faster than powers.

(iv)   This is an example which does not easily fit into the standard patterns established in Proposition 3. Both components of the subtraction tend to infinity and so there is no algebraic way of guessing the limit. We should not try to treat infinity as a number, and an argument which says that since both parts tend to infinity the sequence overall must tend to $\infty - \infty = 0$ is invalid, and in fact the conclusion is false as we shall see. You are encouraged to explore values of this sequence on your calculator to see whether you can guess what the limit might be. We shall in fact establish it algebraically by rationalizing the difference of two square roots, using the factorization $a^2 - b^2 = (a - b)(a + b)$ as follows

$$\sqrt{n^2 + n} - n = \frac{(\sqrt{n^2 + n} - n)(\sqrt{n^2 + n} + n)}{(\sqrt{n^2 + n} + n)} = \frac{n}{\sqrt{n^2 + n} + n}$$

$$= \frac{1}{\sqrt{1 + \frac{1}{n}} + 1} \to \frac{1}{1 + 1} = \frac{1}{2}.$$

Note that we have assumed that if $\lim(a_n) = l > 0$ then $\lim \sqrt{a_n} = \sqrt{l}$.

TUTORIAL PROBLEM 8

Try to construct a proof that if $\lim(a_n) = l > 0$ then $\lim \sqrt{a_n} = \sqrt{l}$. For what sorts of other functions f do you think that $a_n \to l$ implies that $f(a_n) \to f(l)$?

TUTORIAL PROBLEM 9

Adapt the method used in proving Proposition 3(iii) to show that $\lim(n^{1/\sqrt{n}}) = 1$. You will need to use the binomial expansion and pick out the term containing t_n^3, and then rearrange the resulting inequality. Try to generalize this result, replacing $\sqrt{n}$ by other powers of n.

EXERCISES 7.3

1. A sequence (b_n) is defined in terms of the sequence (a_n) by $b_n = |a_n|$. Prove that if $\lim(a_n) = l$ then $\lim(b_n) = |l|$. Give examples to show that the converse is not necessarily true.

2. Prove that if $a_n \to \infty$ then $\lim(1/a_n) = 0$, using an approach similar in structure to that of Example 4.

3. Determine the limiting behaviour of the sequences whose nth terms are given by the following formulae.

   (i) $1 + \dfrac{1}{n^2}$,   (ii) $n^2 - 2n$,   (iii) $n(1 + (-1)^n)$,   (iv) $\dfrac{n!}{n^n}$,   (v) $\dfrac{a^n}{b^n}$,

   (vi) $\dfrac{a^n}{n!}$,   (vii) $\dfrac{n^2 \ln(n)}{(n+1)2^n}$,   (viii) $\sqrt[n]{n + \sqrt{n}}$,   (ix) $\dfrac{n^5}{2^n}$,

   (x) $\dfrac{2^n}{\sqrt{(n^5 - 2n)}}$,   (xi) $\dfrac{n^2}{\sqrt{(n^3 - 2)}}$,   (xii) $\sqrt[n]{(n^2)}$,   (xiii) $\dfrac{n + 3^{n-1}}{n^2 + 2^n + 3^n}$,

   (xiv) $\dfrac{\sqrt{n+1} - \sqrt{n}}{\sqrt{n} - \sqrt{n-1}}$,   (xv) $\sqrt[n]{a^n + b^n}$ $(a > b > 0)$,   (xvi) $\dfrac{10^n}{n^{10}}$,

   (xvii) $\sin^2\left(\dfrac{n\pi}{4}\right)$,   (xviii) $n\cos\left(\dfrac{1}{n}\right)$,   (xix) $\sin^n \theta$,   (xx) $\cos(n\theta)$.

4. Give examples of sequences $(a_n), (b_n)$ such that $a_n \to \infty$ and $b_n \to \infty$ while

   (i) $(a_n - b_n) \to -\infty$,   (ii) $(a_n - b_n) \to 0$,   (iii) $(a_n - b_n) \to \infty$,

   (iv) $(a_n - b_n) \to 2$,   (v) $(a_n - b_n)$ oscillates unboundedly,

   (vi) $\left(\dfrac{a_n}{b_n}\right) \to \infty$,   (vii) $\left(\dfrac{a_n}{b_n}\right) \to 0$,   (viii) $\left(\dfrac{a_n}{b_n}\right) \to 3$.

7.4 Increasing and Decreasing Sequences

Some of the sequences we have encountered have behaved in a regular fashion in the sense that the successive terms have increased in value, for example

$$1, 3, 5, 7, 9, \ldots, \quad \text{or} \quad \frac{1}{2}, \frac{2}{3}, \frac{3}{4}, \frac{4}{5}, \frac{5}{6}, \ldots$$

In the first case the sequence has no limit, whereas in the second case the limit will be 1. Even though the behaviour is clear, we shall formulate a definition.

● *Definition 4*

   (i) A sequence (a_n) is said to be non-decreasing if $a_n \leq a_{n+1}$ for all $n \in \mathbb{N}$. (Some books use the term 'weakly increasing' instead of 'non-decreasing'.)

   (ii) A sequence (a_n) is said to be strictly increasing if $a_n < a_{n+1}$ for all $n \in \mathbb{N}$.

   (iii) A sequence (a_n) is said to be non-increasing if $a_n \geq a_{n+1}$ for all $n \in \mathbb{N}$.

   (iv) A sequence (a_n) is said to be strictly decreasing if $a_n > a_{n+1}$ for all $n \in \mathbb{N}$.

(v) A sequence satisfying any of these four criteria is said to be monotonic (the word signifies that the terms of the sequence move in one direction—up or down in value). So a sequence which is neither increasing nor decreasing can simply be said to be not monotonic.

The distinction between (i) and (ii) is exemplified by a sequence such as $1, 1, 2, 2, 3, 3, \ldots$, which satisfies (i) but not (ii). In this section the results we obtain will cover (i) as well as (ii), although most of the examples will be strictly increasing. We shall use the single word 'increasing' for sequences of type (i) or (ii). Results concerning increasing sequences will have analogues for decreasing sequences. In most cases the relationship can be established simply by noting that if (a_n) is a decreasing sequence then the sequence $(-a_n)$ is increasing. We shall therefore concentrate on increasing sequences as far as establishing proofs are concerned. A significant difference between the first two sequences quoted in this section is that the terms of the first can become arbitrarily large, whereas those of the second cannot. The terms of the second form a bounded set in the sense of Definition 1 of Chapter 5. We shall repeat that definition here, in the context of sequences, to save continual reference back to Chapter 5. ●

● Definition 5

(i) A sequence (a_n) is said to be bounded above if there is a real number K such that $a_n \leq K$ for all $n \in \mathbb{N}$, i.e. all the terms of the sequence are less than (or equal to) K.

(ii) A sequence (a_n) is said to be bounded below if there is a real number H such that $a_n \geq H$ for all $n \in \mathbb{N}$, i.e. all the terms of the sequence are greater than (or equal to) H.

(iii) A sequence is said to be bounded if it is bounded above and bounded below.

(iv) A sequence is unbounded if it does not satisfy (iii), i.e. if it is not bounded above or if it is not bounded below (or both). ●

TUTORIAL PROBLEM 10

Formulate and prove a version of Proposition 2 in Chapter 5 for sequences.

We shall now consider sequences falling into the various categories specified by Definitions 4 and 5, and investigate their limits. For example, the first sequence in this section is increasing, unbounded and divergent (has no limit). The second is increasing, bounded above and convergent.

TUTORIAL PROBLEM 11

Find examples of sequences to fill as many as possible of the triangular cells in the grid of Fig 7.4. So for example the sequence defined by $a_n = n^2$ would fit into the cell corresponding to increasing, not bounded, and divergent (D).

	Bounded Above		Bounded Below		Not Bounded	
Increasing	C	D	C	D	C	D
Decreasing	C	D	C	D	C	D
Not Monotonic	C	D	C	D	C	D

Fig 7.4 Sequence grid for Tutorial Problem 11.

Having investigated Tutorial Problem 11, it will have been found that some cells cannot be filled. For instance no example will fit the cell corresponding to increasing, bounded above and divergent. This leads to the formulation of the following important result.

● Proposition 4

An increasing sequence (a_n) which is bounded above is convergent, i.e. it has a limit.

●

PROOF

We shall give a proof using the axiom of completeness for the real number system given in Chapter 5. In fact this result is equivalent to that axiom, although we shall not prove that here. In some accounts of the theory of sequences the result is taken as an axiom. This means that the theory of sequences could be explored without having to study the foundations of the real number system first. In fact it will be seen that the definition of least upper bound is structurally similar to the definition of limit. The definitions of bounded sets and bounded sequences are identical.

The set of numbers $S = \{a_n : n \in \mathbb{N}\}$ (the set of terms of the sequence) is a set of numbers which is bounded above. The axiom of completeness tells us that this set has a least upper bound l. We shall prove that $\lim(a_n) = l$.

Let ϵ denote an arbitrary positive number. Using part (ii) of Definition 4 in Chapter 5, with β replaced by ϵ, tells us that $l - \epsilon$ is not an upper bound for the set S and so there is a member of S greater that $l - \epsilon$. This member is a term of the sequence, say a_N. Now the sequence is increasing, so for all $n \geq N$, $a_n \geq a_N$. Also, the set is bounded above by l so that $a_n \leq l$ for all n. This tells us that $0 \leq a_n - l < \epsilon$ for all $n \geq N$. If $0 \leq a_n - l < \epsilon$ then $|a_n - l| < \epsilon$, proving that $\lim(a_n) = l$.

●

TUTORIAL PROBLEM 12

> Formulate and prove a corresponding result for decreasing sequences which are bounded below.

Example 7: The exponential limit

Show that the sequence defined as follows has a limit

$$a_n = \left(1 + \frac{1}{n}\right)^n.$$

This is a sequence for which intuition is not a good guide. It is possible to argue as follows.

(a) The number in brackets is bigger than 1 and it is raised to the power n. So as n increases it will tend to infinity.

(b) The number in brackets has limit 1, and 1 raised to any power is equal to 1, so the limit is 1.

Neither result is correct as we shall see. We cannot separate the roles of n in this way. They are both changing together. Explore this sequence on your calculator.

If we expand a_n by the binomial theorem we obtain

$$a_n = 1 + n.\frac{1}{n} + \frac{n(n-1)}{2!}\frac{1}{n^2} + \ldots + \frac{n(n-1)(n-2)\ldots(n-n+1)}{n!}\frac{1}{n^n}$$

$$= 1 + 1 + \frac{1}{2!}\left(1 - \frac{1}{n}\right) + \ldots + \frac{1}{n!}\left(1 - \frac{1}{n}\right)\left(1 - \frac{2}{n}\right)\ldots\left(1 - \frac{n-1}{n}\right).$$

The $(p+1)$st component in this expression is equal to

$$\frac{1}{p!}\left(1 - \frac{1}{n}\right)\left(1 - \frac{2}{n}\right)\ldots\left(1 - \frac{p-1}{n}\right).$$

This is positive, and for each value of p it increases as n increases. Also, the number of components in the expansion of a_n increases as n increases. Hence (a_n) is an increasing sequence. We can also deduce from the expansion that

$$1 + 1 < a_n < 1 + 1 + \frac{1}{2!} + \frac{1}{3!} \ldots + \frac{1}{n!}$$

$$< 1 + 1 + \frac{1}{2} + \frac{1}{2^2} + \ldots + \frac{1}{2^{n-1}} \quad \text{(G.P.)}$$

$$= 1 + 2.\left(1 - \frac{1}{2^n}\right) < 3.$$

So (a_n) is increasing and bounded above. It therefore has a limit, which the calculations above tell us lies between 2 and 3. In fact the limit is the well-known exponential number $e = 2.71828\ldots$, although we shall not prove that here.

It is in fact the case that for any real number x,

$$\lim\left(1+\frac{x}{n}\right)^n = e^x.$$

This result is sufficiently important to be committed to memory, but is quite complicated to prove.

7.5 Iteration

In this section we return to the procedure of defining sequences by iteration, which we discussed in §7.2. Here we shall develop further the graphical illustration of such sequences and demonstrate the use of Proposition 4 to establish the existence of limits for iterative sequences.

In Fig 7.5 we see graphs of $y = F(x)$ and $y = x$, which we use to solve the equation $x = F(x)$. Suppose that we guess a first approximation a_1, and mark this value on the x-axis. A line drawn from this point parallel to the y-axis will meet the curve at the point $(a_1, F(a_1))$, whose y coordinate is $F(a_1)$, i.e. a_2. If we draw a line parallel to the x-axis from the point on the graph, we meet the y axis at the value a_2 as shown on the figure. We can now copy this value down onto the x-axis by drawing along to the line $y = x$ and then dropping the perpendicular onto the x-axis. We now have the value a_2 on the x-axis and so we can repeat the whole procedure, leading to the value a_3 on the y-axis and thence on the x-axis. On the figure we have continued the procedure beyond this stage, but omitted the parts of the construction going to the axes. Without those we see a kind of stepped line between the two graphs tending towards the point of intersection. This is often referred to as a staircase diagram, and indicates that in this case the sequence is an increasing one. The diagram also suggests that the sequence (a_n) is converging to the limit l found from the solution of $x = F(x)$ at the point of intersection shown in the figure. When we have a particular

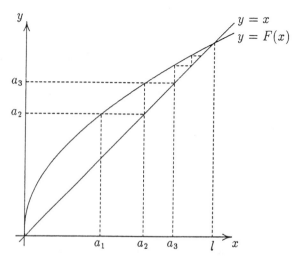

Fig 7.5 Graphical representation for $a_{n+1} = F(a_n)$.

formula for $F(x)$ we can then set about calculating l and then proving analytically that the sequence generated does indeed converge to l, as in the examples below.

It is important to realize that the last step is necessary, as the following example illustrates. Suppose $F(x) = -x$ and that $a_1 = 3$. The solution of $x = F(x)$ is given by $x = -x$, i.e. $x = 0$. But the sequence generated is $3, -3, 3, -3, 3, \ldots$ which does not have a limit. Logically the relationship is that *if* (a_n) has a limit l then $l = F(l)$. This example shows that the converse is not true in general.

Example 8

The sequence (a_n) is defined by

$$a_1 = 1 \quad \text{and} \quad a_n = \frac{3a_{n-1} + 4}{2a_{n-1} + 3}.$$

Show that the sequence has a limit and find the value of the limit.

As in many places in this book, we explore this example from three perspectives, algebraic, graphical and numerical.

We represent the sequence graphically in Fig 7.6, as described above. In order to exhibit the 'staircase' more clearly we have started not with the first term equal to 1, but -1.4. The third term is then 1, but starting the diagram there would give little indication of the pattern of steps because the slope of the curve is small at the point of intersection. The procedure of 'backtracking' from a given point, i.e. given a_n, to find its predecessor according to the iterative formula, was used to choose -1.4 as the starting position for the diagram. Moving backwards like this is important in some applications of iterative sequences, although we shall not encounter them in this book.

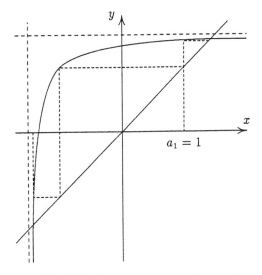

Fig 7.6 Staircase diagram for Example 8.

The diagram suggests that the limit is the positive solution of the equation

$$x = \frac{3x + 4}{2x + 3}, \quad \text{i.e.} \quad x = \sqrt{2}.$$

We can see from the figure that the sequence appears to be increasing and bounded above, with all the terms less than $\sqrt{2}$. We prove these properties algebraically from the formula.

We first prove by induction that $0 < a_n < \sqrt{2}$ for every n. The result is true for $n = 1$ since $a_1 = 1$. We can rewrite the iteration formula as

$$a_n = \frac{3}{2} - \frac{1}{2} \frac{1}{2a_{n-1} + 3}.$$

So if the result is true for a_{n-1}, i.e. $0 < a_{n-1} < \sqrt{2}$, then the original form of the formula makes it clear that $a_n > 0$, and the rearranged form enables us to deduce that

$$a_n = \frac{3}{2} - \frac{1}{2} \frac{1}{2a_{n-1} + 3} < \frac{3}{2} - \frac{1}{2} \frac{1}{2\sqrt{2} + 3} = \sqrt{2}.$$

This gives the result for a_n, and so it is true for all n by induction. This shows that the sequence is bounded above (and below). We now use the result in the form $a_n^2 < 2$ to deduce that for all n

$$a_{n+1} - a_n = \frac{3a_n + 4}{2a_n + 3} - a_n = \frac{4 - 2a_n^2}{2a_n + 3} > \frac{4 - 2 \times 2}{2a_n + 3} = 0.$$

So we have established algebraically that the sequence is increasing and bounded above. It therefore has a limit l. By the algebra of limits

$$\lim \left(\frac{3a_n + 4}{2a_n + 3} \right) = \frac{3l + 4}{2l + 3}.$$

Now $\lim(a_{n+1}) = l$ (see Tutorial Problem 13 below) and so using the iteration formula we conclude that

$$\frac{3l + 4}{2l + 3} = l \quad \text{which gives} \quad l^2 = 2 \quad \text{i.e.} \quad l = \pm\sqrt{2}.$$

All the terms of the sequence are positive, so l cannot be negative. Therefore $l = \sqrt{2}$. We can take this further and analyse the approximation errors, letting $e_n = \sqrt{2} - a_n$. Using the iteration formula gives

$$e_{n+1} = \sqrt{2} - a_{n+1} = \sqrt{2} - \frac{3a_n + 4}{2a_n + 3} = \sqrt{2} - \frac{3(\sqrt{2} - e_n) + 4}{2(\sqrt{2} - e_n) + 3} = \frac{(3 - 2\sqrt{2})e_n}{2(\sqrt{2} - e_n + 3)}.$$

So $\dfrac{e_{n+1}}{e_n} = \dfrac{(3 - 2\sqrt{2})}{2(\sqrt{2} - e_n + 3)} \approx \dfrac{3 - 2\sqrt{2}}{2\sqrt{2} + 3} \approx 0.0295.$

(The first of the approximations above is valid when e_n is small, and in fact is very good after only around five steps.) This shows that the errors decrease quite rapidly, so this is an iteration with a good rate of convergence. Evaluating the first few terms of the sequence to seven places of decimals, using a calculator, gives, starting with $a_1 = 1$,

$$a_2 = 1.4 \qquad a_3 = 1.4137931$$
$$a_4 = 1.4142011 \qquad a_5 = 1.4142131$$
$$a_6 = 1.4142135 \qquad a_7 = 1.4142135$$

The errors for a_6 and a_7 are approximately 3.036×10^{-8} and 9.2×10^{-10} respectively. This fits quite well with the estimate above for the ratio of successive errors.

TUTORIAL PROBLEM 13

Explain why $\lim(a_n) = l$ implies $\lim(a_{n+1}) = l$.

Example 9

The sequence (a_n) is defined by

$$a_1 = 2 \quad \text{and} \quad a_n = \frac{1}{a_{n-1} + 2}.$$

Show that the sequence has a limit and find the value of the limit.

A few iterations on a calculator will begin to clarify the behaviour of this sequence, giving (to seven decimal places),

$$a_1 = 2 \qquad a_2 = 0.25$$
$$a_3 = 0.4444444 \qquad a_4 = 0.4090909$$
$$a_5 = 0.4150943 \qquad a_6 = 0.4140625$$
$$a_7 = 0.4142394 \qquad a_8 = 0.4142091$$
$$a_9 = 0.4142143 \qquad a_{10} = 0.4142134$$

This confirms that the odd terms form a decreasing sequence and the even terms form an increasing sequence. The graph and the data suggest that the odd terms are all bigger than the intersection value $\sqrt{2} - 1$ and that the even terms are all less than this value. We shall prove all these facts algebraically.

In Fig 7.7 we have drawn the graphs of $y = x$ and $y = 1/(x+2)$. The formula tells us that all the terms of the sequence will be positive, and so it is the positive intersection which is relevant. The two graphs intersect where $x = 1/(x+2)$, giving $x = \sqrt{2} - 1$. We have begun to plot the sequence using the graphs as explained at the beginning of this section. In the window to the right of the graph we have magnified schematically the part of the graph near to the intersection, in order to exhibit the behaviour of the sequence more clearly. (We can use a Zoom facility which Graphical Calculus has, and which is also on some graphics calculators.) This time we do not see a staircase but a plot which winds around the point of intersection. For obvious reasons this type of configuration is referred to as a cobweb diagram. The visual properties reflect the numerical calculations above.

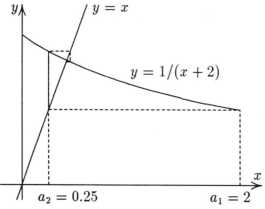
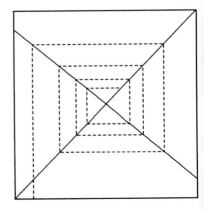

Fig 7.7 Cobweb diagram for Example 9.

Firstly $a_1 = 2 > \sqrt{2} - 1$. If $a_{n-1} > \sqrt{2} - 1$ then

$$0 < a_n = \frac{1}{a_{n-1} + 2} < \frac{1}{(\sqrt{2} - 1) + 2} = \frac{1}{\sqrt{2} + 1} = \sqrt{2} - 1.$$

Secondly if $0 < a_{n-1} < \sqrt{2} - 1$ then

$$a_n = \frac{1}{a_{n-1} + 2} > \frac{1}{(\sqrt{2} - 1) + 2} = \frac{1}{\sqrt{2} + 1} = \sqrt{2} - 1.$$

This shows that the terms of the sequence alternate either side of $\sqrt{2} - 1$.

In order to deal with the even and odd terms separately we need to relate a_n to a_{n+2}.

$$a_{n+2} = \frac{1}{a_{n+1} + 2} = \frac{1}{\dfrac{1}{a_n + 2} + 2} = \frac{a_n + 2}{2a_n + 5}.$$

We can now use this to investigate the relative sizes of a_n and a_{n+2} as follows

$$a_{n+2} - a_n = \frac{a_n + 2}{2a_n + 5} - a_n = \frac{a_n + 2 - 2a_n^2 - 5a_n}{2a_n + 5}$$

$$= \frac{2(1 - 2a_n - a_n^2)}{2a_n + 5} = \frac{2((\sqrt{2} - 1) - a_n)((\sqrt{2} + 1) + a_n)}{2a_n + 5}.$$

Considering the sign of the first bracket in the last numerator tells us that if $a_n > \sqrt{2} - 1$ then $a_{n+2} - a_n < 0$. It also tells us that if $0 < a_n < \sqrt{2} - 1$ then $a_{n+2} - a_n > 0$.

An alternative, and often useful, method of dealing with $a_{n+2} - a_n$ is to substitute for *both* terms. Omitting the details of the algebra, this gives

$$a_{n+2} - a_n = \frac{a_n + 2}{2a_n + 5} - \frac{a_{n-2} + 2}{2a_{n-2} + 5} = \frac{a_n - a_{n-2}}{(2a_n + 5)(2a_{n-2} + 5)}.$$

So the sign of $a_{n+2} - a_n$ is the same as the sign of $a_n - a_{n-2}$. From the calculations at the beginning of this example we know that $a_3 - a_1 < 0$ and $a_4 - a_2 > 0$. This enables us to deduce that $(a_1, a_3, a_5, \ldots)$ is a decreasing sequence, and that $(a_2, a_4, a_6, \ldots)$ is an increasing sequence.

Putting these results together, we have shown that when $a_1 = 2$ the sequence $(a_1, a_3, a_5, \ldots)$ is decreasing and bounded below, and that the sequence $(a_2, a_4, a_6, \ldots)$ is increasing and bounded above. So both sequences have limits and, from the equation relating a_{n+2} to a_n, it follows that both limits will satisfy $l = (l+2)/(2l+5)$. Solving this equation gives $l = \sqrt{2} - 1$ or $l = -(\sqrt{2}+1)$. All the terms are positive so the first solution is the one which applies to both the even and odd subsequences. So both subsequences have the same limit, and therefore the entire sequence converges to $\sqrt{2} - 1$.

EXERCISES 7.5

1. Draw the curves $3y = x^3 + 2$ and $y = x$ on the same axes. Use the graphs to illustrate the fact that if $-2 < a_1 \leq 1$ then the sequence (a_n) defined by $3a_n = a_{n-1}^3 + 2$ converges to 1. Prove this result, showing that for a_1 in the stated interval the sequence is increasing and bounded above by 1. If $e_n = 1 - a_n$ obtain an estimate of the ratio e_{n+1}/e_n. What does this tell you about the rate of convergence of the sequence (a_n)? Check with a calculator.

    Find out what happens if a_1 lies outside the interval between -2 and 1, and again prove your results algebraically.

2. Investigate the following sequences in the same way as in Exercise 1, considering a_1 in the various intervals specified.

    (i) $\quad a_n = \dfrac{1}{a_{n-1}} + 2, \quad (a_1 > 0; a_1 < 0),$

    (ii) $\quad a_n = 2 - \dfrac{1}{a_{n-1}}, \quad (a_1 > 1; a_1 < 0; 0 < a_1 < 1/2; 1/2 < a_1 < 1).$

3. Express the terms a_n of the sequence

    $$\sqrt{2}, \sqrt{2 + \sqrt{2}}, \sqrt{2 + \sqrt{2 + \sqrt{2}}}, \ldots$$

    by means of a relationship expressing a_n as a function $F(a_{n-1})$ of a_{n-1}. Illustrate graphically how the terms a_n increase towards 2. Prove by induction from your iterative formula that $\sqrt{2} \leq a_n < 2$ for all n and that $a_{n+1} > a_n$ for all n. Deduce that a_n has a limit and find its value. Prove also that $|a_{n+1} - 2| < |a_n - 2|/2$ for all n. What does this tell you about the rate of convergence? Check the results on a calculator.

7.6 Complex Sequences

If we consider Definition 2 for the limit of a sequence we see that it involves the quantity $|a_n - l|$. We can therefore interpret this as the modulus and thereby apply the definition to sequences of complex numbers. Recall from Chapter 6 that $|a_n - l|$

gives the distance between the complex numbers a_n and l in the complex plane, and so to say that $|a_n - l| < \epsilon$ says that a_n is within distance ϵ of l. In the complex plane this means that a_n lies within a circle with centre l and radius ϵ. Thinking of ϵ as being small says that a_n and l are very close together in the complex plane. This fits in with our intuitive idea of the limit of a sequence where we think of the terms getting closer and closer to the limit. Since the definition is the same for complex as for real sequences it follows that consequences such as Proposition 1 on the algebra of limits are also valid, with identical proofs. Results (iv) and (v) of Proposition 3 also carry across to complex sequences. It is worth checking the proofs to confirm that x could represent a complex number in those contexts. We can relate the behaviour of complex sequences to the sequences of real and imaginary parts as follows.

● *Proposition 5*

Suppose (a_n) is a sequence of complex numbers. Then

$$\lim(a_n) = l \quad \text{if and only if} \quad \lim(\text{Re}(a_n)) = \text{Re}(l) \quad \text{and} \quad \lim(\text{Im}(a_n)) = \text{Im}(l) \quad ●$$

PROOF
We shall exhibit the geometrical reasoning which leads to verification of the limit definitions. We shall leave the interested reader to write out a complete analytical proof, perhaps with tutorial help.

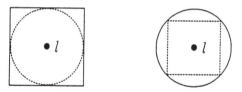

Fig 7.8 Regions for convergence of complex sequences.

If $\text{Re}(a_n - l) < \epsilon/\sqrt{2}$ and $\text{Im}(a_n - l) < \epsilon\sqrt{2}$ then a_n lies in the square with centre l and sides of length $2\epsilon/\sqrt{2}$ parallel to the real and imaginary axes. The right-hand diagram in Fig 7.8 shows that a_n will then be in the circle with centre l and radius ϵ, so that $|a_n - l| < \epsilon$. This deduction is used to show that if $\lim(\text{Re}(a_n)) = \text{Re}(l)$ and $\lim(\text{Im}(a_n)) = \text{Im}(l)$ then $\lim(a_n) = l$.

If $|a_n - l| < \epsilon$ then the left-hand diagram shows that a_n will lie within the square with centre l and sides of length 2ϵ, so that $\text{Re}(a_n - l) < \epsilon$ and $\text{Im}(a_n - l) < \epsilon$. This deduction is used to show that if $\lim(a_n) = l$ then $\lim(\text{Re}(a_n)) = \text{Re}(l)$ and $\lim(\text{Im}(a_n)) = \text{Im}(l)$.　　　　　　　　　●

TUTORIAL PROBLEM 14

Investigate the behaviour of the sequence $(e^{in\theta})$ for various values of θ.

EXERCISES 7.6

1. Show that for a complex sequence (a_n), $\lim(a_n) = 0$ if and only if $\lim(|a_n|) = 0$. Deduce that if $a_n = r_n(\cos(\theta_n) + \mathrm{i}\sin(\theta_n))$ then $\lim(a_n) = 0$ if and only if $\lim(r_n) = 0$.

2. Determine the limiting behaviour of the sequences (a_n) defined as follows.

   (i) $a_n = (0.7 + 0.7\mathrm{i})^n$, (ii) $a_n = (0.8 + 0.8\mathrm{i})^n$,

   (iii) $a_n = (0.8 + 0.6\mathrm{i})^n$, (iv) $a_n = \mathrm{e}^{\mathrm{i}\alpha/n}$.

   If you have computer graphics facilities (as with Basic on a BBC machine, or TurboPascal, or a programmable graphics calculator) write programs to plot some terms of these sequences in the complex plane (in (iv) you will have to choose some specific values to try for α).

Summary

In this chapter we have considered infinite sequences of numbers. The idea of a sequence which can continue without coming to an end is related to mathematical induction, particularly by using induction to define terms of a sequence from their predecessors. This leads to the idea of iteration as a method of generating a sequence, especially in the context of successive approximations.

In seeking to give a mathematical definition of the limiting behaviour of sequences we have relied on a detailed introductory example, but also, naturally, on the historical development of the formulation of Definitions 2 and 3. The fact that these took eminent European mathematicians much of the 18th and 19th centuries to accomplish is testimony to the logical complexities involved. This section of the chapter is therefore the most demanding in the book. Proposition 1, on the algebra of limits, shows these definitions in action, being used to prove results we apply all the time. The proofs should be studied with a view to gaining some understanding of how the logic of the definitions is handled, and will benefit from tutorial discussion. In the author's view it may not be appropriate to try to learn the proofs at this stage. Proposition 3 on the other hand shows the definitions in action in a less abstract setting, dealing with particular sequences. The purpose of that proposition is to establish a small number of standard results which will be used frequently, in conjunction with the rules for limits embodied in Proposition 1. This is very similar to the methods involved in differentiation and integration, where one learns a few standard results and then applies rules for products, quotients etc. The methods of this section are important and should be mastered through the examples and exercises.

We have considered the relationship between the existence of a limit for a sequence and the completeness of the real number system studied in Chapter 5. These ideas were closely interrelated historically. The establishment of the exponential limit is one consequence, and the result, in Example 7, should be learned. The main application is to iterative procedures, and here I have re-emphasized the importance of multiple perspectives, with algebraic, graphical and numerical aspects studied. Omitting one of these systematically will certainly impede the development of a clear understanding of the behaviour of sequences, and computers and calculators can greatly assist with numerical and graphical situations.

Finally we have considered very briefly sequences of complex numbers, again emphasizing the geometrical view, given through the complex plane.

EXERCISES ON CHAPTER 7

1. Rearrange the equation $e^x = 3x^2 - 2$ in the form $x = F(x)$ in as many ways as you can think of. Use a graphics calculator or a computer package to draw graphs of $y = F(x)$ and $y = x$. Investigate the behaviour of the associated sequences given by $a_n = F(a_{n-1})$, choosing various initial values.

2. Assuming that if $\lim(a_n) = l, (a_n > 0, l > 0)$ then $\lim(\ln(a_n)) = \ln(l)$, use the properties of the logarithmic function together with the result of Proposition 3(iii) to show that

$$\lim\left(\frac{\ln(n)}{n}\right) = 0.$$

Use the results from Tutorial Problem 9 to show that replacing n by a positive power of n in the denominator of $\ln(n)/n$ still gives limit zero. This tells us that powers tend to infinity faster than logarithms.

3. Determine the limiting behaviour of the sequences whose nth terms are given by the following formulae.

(i) $\dfrac{n^3 - 3n^2 + 2}{3(n-1)^3 + 4}$, (ii) $\left(1 + \dfrac{1}{n}\right)^{n/2}$, (iii) $\dfrac{\sin^3 n}{2n^2}$, (iv) $n^{n/(n^2+1)}$,

(v) $(\sqrt[n]{n} - 1)^n$, (vi) $\dfrac{n^3 + \ln n + 2^n}{n^2 + \sqrt{n} + 3^n}$, (vii) $\dfrac{n^2 + 1}{1 - 2n - n^2}$,

(viii) $\dfrac{\sin(e^n)}{n}$, (ix) $\dfrac{n(1 + (-1)^n)}{\ln n}$.

4. The sequence (a_n) is defined as follows. Prove that the sequence is increasing and bounded above, and therefore converges.

$$a_n = \frac{1}{n+1} + \frac{1}{n+2} + \ldots + \frac{1}{2n}.$$

5. The sequence (a_n) is defined by $a_1 = 6$ and

$$a_n = \frac{11a_{n-1} - 8}{a_{n-1} + 5}.$$

Illustrate the behaviour of this sequence graphically. Prove from the recurrence equation that $a_n > 4$ for all n, and that a_n is a decreasing sequence. Deduce that the sequence has a limit, and find its value.

6. Let S denote an infinite set of real numbers, bounded above, with least upper bound l. Use the definition of l.u.b.(S) to show that it is possible to find a strictly increasing sequence (a_n) of numbers belonging to S for which $\lim(a_n) = l$.

8 • Infinite Series

Infinite series are very important in many branches of mathematics, and like sequences are used to provide approximations. In particular, they are used to provide polynomial approximations to trigonometric and many other classes of functions, and also to enable us to find out about functions for which there is no algebraic formula. Many mathematicians have investigated their properties, and with the invention of calculus they achieved great prominence. Like so many topics we find the name of Euler associated with discoveries involving series.

The Ancient Greek philosopher Zeno (c. 450 BC) was responsible for expounding a number of so-called 'paradoxes' concerned with motion. One of them concerned Achilles, who could run very swiftly, and a tortoise. An account of this paradox using measures of distance is as follows. Suppose that Achilles races against the tortoise and starts 100 metres behind. Suppose that Achilles can run 10 times faster than the tortoise. After Achilles has covered 100 metres the tortoise will still be 10 metres in front. After Achilles has covered those 10 metres the tortoise will still be leading by 1 metre. When Achilles has covered that one metre distance the tortoise will have moved on 10 cm. This process can continue for ever, so that Achilles will never catch the tortoise.

Achilles   Tortoise

Fig 8.1 Achilles chasing the tortoise.

We leave it to readers to discuss a resolution of the paradox itself, but we shall consider a method of finding where Achilles catches the tortoise. We can use the reasoning above to say that Achilles has to cover all the distances specified, continuing potentially without end, and so the total distance will in some sense be the sum of these distances, which in metres will therefore be

$$100 + 10 + 1 + 0.1 + 0.01 + 0.001 + \ldots \text{(continuing without stopping).}$$

We have no way of performing such a calculation by the ordinary processes of arithmetic, but we can find the total distance which Achilles has covered at each stage of the procedure. After the first stage he has run 100 m, after the second stage 110 m, after the third stage 111 m, and so on. This gives us a potentially infinite sequence of total distances beginning

$$100, 110, 111, 111.1, 111.11, 111.111, \ldots.$$

We can see that all these distances are less than 112 m, and that the sequence is increasing. This tells us, by Proposition 4 of Chapter 7, that the sequence has a limit. This limit would seem to be a sensible interpretation for the potentially infinite addition of distances. What we have done here is to start with an infinite sequence of separate distances and generated an infinite sequence of partial totals. The limit of the sequence of totals suggests itself as a reasonable definition of the infinite sum of the separate distances.

TUTORIAL PROBLEM I

Find an iterative formula for the total distance d_n covered after n stages of Achilles' pursuit of the tortoise, where $d_1 = 100$. Investigate the convergence of this sequence by the methods of §7.5.

8.1 Convergence

In the introduction we considered the process of 'infinite addition' by generating a sequence of partial sums. This close relationship contributes to a confusion about the difference between a sequence and a series. After all, the words 'convergence' and 'divergence' are used in both contexts. An additional factor is the use of the words in everyday language. We often encounter things like 'find the next letter in the *series* J, F, M, A, M, ...', and indeed in some school textbooks the word 'series' is used in place of 'sequence' for a succession of letters or numbers. The description of the process in the introduction suggests the formal Definition 1 below, which makes the distinction clear, in that a series is defined as a *pair* of sequences related in a particular way. The confusion can also arise because textbooks very rarely use this formal definition in discussion about series, but always use the associated Σ notation. In fact, this practice has some advantages, and we shall follow it in this book.

● *Definition I*

An infinite series is an ordered pair $((a_n), (s_n))$ of infinite sequences, where $s_1 = a_1$ and $s_{n+1} = s_n + a_{n+1}$. a_n is called the nth term of the series and s_n is called the nth partial sum of the series. If the sequence (s_n) of partial sums has a limit then the series is said to converge, and the value of $\lim(s_n)$ is called the sum of the infinite series. Otherwise the series is said to diverge. ●

Notice that we have defined the partial sums from the terms, and this is the order relating to most situations involving series. We can invert the definition if we wish by starting with a sequence (s_n) and then defining the sequence (a_n) by $a_1 = s_1$ and

$a_n = s_n - s_{n-1}$. Notice also that the definition works equally well with real or complex terms, although a majority of the examples in this chapter involve real numbers.

TUTORIAL PROBLEM 2

> Show that if we start with s_n and define a_n as above, then applying Definition 1 to the resulting sequence (a_n) yields the sequence (s_n) we started with, so that the two approaches are indeed equivalent.

We now introduce some further notation and terminology. Firstly if we look back to the definition of the Σ notation in §2.1 we see that the structure is exactly that of Definition 1, and so using that notation we have

$$s_n = \sum_{i=1}^{n} a_i.$$

When the series converges, so that $\lim(s_n) = S$, say, we write

$$S = \sum_{i=1}^{\infty} a_i \quad \text{or} \quad S = \sum_{i \in \mathbb{N}} a_i.$$

The latter notation avoids the danger of interpreting the symbol ∞ as if it were a number. However, it is much less commonly used than the former in textbooks, and so we shall use the first notation with the 'health warning' concerning ∞. We will often use the even more abbreviated notation Σa_i. Another point about the notation is that it is normally used (ambiguously) to denote both the series and its sum (when convergent). So we find statements like

$$\text{the series} \quad \sum_{i=1}^{\infty} \frac{1}{i^2} = \frac{\pi^2}{6}. \qquad \text{The series} \quad \sum_{i=1}^{\infty} \frac{1}{i} \quad \text{diverges.}$$

We shall begin discussion of examples with what may be a familiar situation, that of a geometric series.

● *Example 1*

Find a formula for the nth partial sum of the series whose nth term is given by $a_n = ar^{n-1}$, where a and r are constants. Discuss convergence of the associated infinite series.

It is often helpful to write out the first few terms of a sequence or a series explicitly, so that in this case the sequence of terms is $(a, ar, ar^2, ar^3, \ldots)$. We see that each term is obtained from its predecessor by multiplying by r, which can also be seen as the invariant ratio of successive terms. So r is called the common ratio, or common multiplier. There is a well-known and rather ingenious method of finding a formula for s_n as follows.

$$s_n = a + ar + ar^2 + ar^3 + \ldots + ar^{n-1},$$

$$rs_n = \quad ar + ar^2 + ar^3 + \ldots + ar^{n-1} + ar^n.$$

So $\quad (1 - r)s_n = a \qquad\qquad\qquad\qquad - ar^n.$

So if $r \neq 1$ we obtain

$$s_n = \frac{a(1 - r^n)}{1 - r}.$$

If $r = 1$ then all the terms are equal to a and so $s_n = na$. Now if $|r| < 1$ then $\lim(r^n) = 0$ and so $\lim(s_n) = a/(1 - r)$. If $r \geq 1$ then $s_n \to \infty$. If $r = -1$ then the sequence s_n is bounded but has no limit. The best way to appreciate this is to write out the first few terms of both (a_n) and (s_n) in this case. If $r < -1$ then the sequence s_n oscillates unboundedly.

Example 2

Find a formula for the nth partial sum of the trigonometric series whose nth term is given by $a_n = \cos(n\theta)$. Discuss convergence of the series.

The easiest method of dealing with this problem is to use the complex exponential (§6.5). This will enable us to sum both the cosine series and the related sine series together. We obtain a geometric series with common ratio $e^{i\theta}$.

$$\sum_{k=1}^{n} (\cos k\theta + i \sin k\theta) = \sum_{k=1}^{n} e^{ik\theta} = e^{i\theta} \sum_{k=1}^{n} e^{i(k-1)\theta} = e^{i\theta} \frac{e^{in\theta} - 1}{e^{i\theta} - 1} \quad (e^{i\theta} \neq 1),$$

using the result of Example 1. (We leave it to the reader to deal with the case $e^{i\theta} = 1$.) We now need to separate this into real and imaginary parts. We use the identity

$$e^{i\alpha} - 1 = e^{i\alpha/2}(e^{i\alpha/2} - e^{-i\alpha/2}) = e^{i\alpha/2} 2i \sin(\alpha/2),$$

(see Exercise 4 of §6.5.) Applying this result to the previous expression gives

$$e^{i\theta} \frac{e^{in\theta} - 1}{e^{i\theta} - 1} = e^{i\theta} \frac{e^{in\theta/2} \, 2i \sin(n\theta/2)}{e^{i\theta/2} \, 2i \sin(\theta/2)} = e^{i(n+1)\theta/2} \frac{\sin(n\theta/2)}{\sin(\theta/2)}.$$

Finally this gives

$$\sum_{k=1}^{n} \cos(k\theta) = \cos\left(\frac{n+1}{2}\theta\right) \frac{\sin(n\theta/2)}{\sin(\theta/2)}$$

and

$$\sum_{k=1}^{n} \sin(k\theta) = \sin\left(\frac{n+1}{2}\theta\right) \frac{\sin(n\theta/2)}{\sin(\theta/2)}.$$

We can see from summing the geometric series that convergence of the infinite series depends on the limiting behaviour of the sequence $(e^{in\theta})$. Now these complex numbers always lie on the unit circle, centred at the origin in the complex plane. If n increases by 1 then the argument increases by θ, so that the sequence of numbers moves around the circle in regular steps. If θ is a rational multiple of π then $e^{i\theta}$ is a root of unity, and so the sequence will repeat, with the terms moving around the vertices of the appropriate regular polygon (see §6.6). The sequence therefore does not have a limit. If θ is not a rational multiple of π then the sequence does not repeat, but the terms are uniformly distributed around the circle, and again do not tend to a limit. This result is not easy to prove and is outside the scope of this book. We shall return later to the question of convergence of this series. Notice that in this example we have used k as the variable of summation rather than i, because we are naturally using i to denote the complex number. Other symbols for this variable could have been chosen, for example j, r, s, etc.

Example 3

Prove that the series whose nth term is $1/n(n+1)$ converges, and find its sum.

As with the previous examples, we can find an explicit formula for the partial sums of this series. We first decompose the term into partial fractions,

$$\frac{1}{n(n+1)} = \frac{1}{n} - \frac{1}{n+1}.$$

The easiest way to see how these add up is to write out the pattern of the first few terms without the Σ notation. So

$$\sum_{i=1}^{n} \frac{1}{i(i+1)} = \sum_{i=1}^{n} \left(\frac{1}{i} - \frac{1}{i+1} \right)$$

$$= \left(\frac{1}{1} - \frac{1}{2} \right) + \left(\frac{1}{2} - \frac{1}{3} \right) + \left(\frac{1}{3} - \frac{1}{4} \right) + \ldots + \left(\frac{1}{n} - \frac{1}{n+1} \right).$$

We can see that most of the fractions cancel, and we are left with

$$\sum_{i=1}^{n} \frac{1}{i(i+1)} = 1 - \frac{1}{n+1}.$$

Now $\lim(1/(n+1)) = 0$, so the series converges, and its sum is 1.

One very important infinite series is the exponential series

$$1 + \frac{1}{1!} + \frac{1}{2!} + \frac{1}{3!} + \ldots$$

which converges to the exponential number $e \approx 2.71828$, which we encountered in Example 7 of Chapter 7. We sometimes see this expressed in Σ notation as

$$e = \sum_{n=0}^{\infty} \frac{1}{n!},$$

i.e. with the summation starting at $n = 0$ rather than $n = 1$, and where the convention that $0! = 1$ is used, as it is in probability theory. A complete proof that the series converges to the exponential number is outside the scope of this book.

EXERCISES 8.1

1. Adapt the method used in Example 1 to find a formula for the partial sum

$$1 + 2r + 3r^2 + 4r^3 + \ldots + nr^{n-1}.$$

   Show that for $|r| < 1$ the associated infinite series converges, and find its sum. Investigate the cases $|r| > 1, r = 1, r = -1$.

2. Use the method of partial fractions in a similar way to that of Example 3 to obtain a formula for the sum of the finite series

$$\sum_{j=1}^{n} \frac{j}{(j+4)(j+5)(j+6)}.$$

   Show that the associated infinite series converges with sum $1/10$.

3. Using the method from Example 2, find formulae for

$$\cos(\alpha + \theta) + \cos(\alpha + 2\theta) + \cos(\alpha + 3\theta) + \ldots + \cos(\alpha + n\theta),$$

$$\cos(\alpha + \theta) - \cos(\alpha + 2\theta) + \cos(\alpha + 3\theta) - \ldots + (-1)^{n+1} \cos(\alpha + n\theta),$$

   and the associated sine series.

8.2 Tests for Convergence

If we can find explicit formulae for partial sums then we can determine convergence as above by considering the sequence of partial sums. In many cases however such a formula cannot be obtained, and so to find the sum we have to invoke some algebraic or analytical theory, or estimate numerically by calculating the partial sum for a large number of terms. Before embarking on such investigations one needs to know whether the series does converge, for otherwise all such approaches will be a waste of effort. Much of the rest of this chapter is concerned with this idea of testing a series for convergence in the absence of partial sum formulae. The first result is, in effect, a test for divergence.

● *Proposition 1*

If the series Σa_n converges then $\lim(a_n) = 0$.   ●

PROOF
Let (s_n) denote the sequence of partial sums, and let S denote the sum of the series, so that $S = \lim(s_n)$. Now $a_{n+1} = s_{n+1} - s_n$ and using the result of Tutorial Problem 13 in Chapter 7 tells us that $\lim(s_{n+1}) = S$. The algebra of limits together with the tutorial problem just referred to then gives

$$\lim(a_n) = \lim(a_{n+1}) = \lim(s_{n+1} - s_n) = \lim(s_{n+1}) - \lim(s_n) = S - S = 0.   ●$$

It is very important to be careful about the logic of this result. Using the logical symbol for implication we can abbreviate the statement to

$$\sum a_n \text{ convergent } \Rightarrow \lim(a_n) = 0.$$

The contrapositive of this can be written

$$\lim(a_n) \neq 0 \Rightarrow \sum a_n \text{ diverges.}$$

Note that in this statement $\lim(a_n) \neq 0$ should really be amplified to include the case where (a_n) does not have a limit. This can now be applied to deal with the convergence part of Example 2. There we have $a_n = e^{in\theta}$, and because $|a_n| = 1$ this means that we cannot have $\lim(a_n) = 0$ and so the geometric series in that example cannot be convergent.

Finally, and most importantly, we should notice that the converse of Proposition 1 is not true, as the next example will show. This means that knowing $\lim(a_n) = 0$ does not enable us to make any deduction about whether Σa_n converges. In the language of §1.3, $\lim(a_n) = 0$ is a necessary but not a sufficient condition for convergence of Σa_n. The use of the converse is a very common error in this context, and should be guarded against. The results of the next two examples should be learned as illustrations of this situation. The series in Example 4 diverges whereas that in Example 5 converges, but in both cases the nth term of the series tends to zero.

Example 4

The series $\Sigma(1/n)$ diverges.

We shall show how the proof works, but we will not include all the algebraic details. We write out a number of terms of the series and group them as shown.

$$1 + \frac{1}{2} + \left(\frac{1}{3} + \frac{1}{4}\right) + \left(\frac{1}{5} + \frac{1}{6} + \frac{1}{7} + \frac{1}{8}\right) + \left(\frac{1}{9} + \frac{1}{10} + \cdots + \frac{1}{16}\right) + \cdots$$

$$> 1 + \frac{1}{2} + \frac{2}{4} + \frac{4}{8} + \frac{8}{16} + \cdots$$

$$= 1 + \frac{1}{2} + \frac{1}{2} + \frac{1}{2} + \frac{1}{2} + \cdots$$

We can clearly continue in this fashion and produce a partial sum with as many halves as we like, so that the sequence of partial sums will be unbounded and have no limit. (Readers with a taste for complete analytical proofs can fill in all the details.)

Example 5

The series $\Sigma(1/n^2)$ converges.

We consider the partial sums of this series, using the partial fraction approach of Example 3.

$$s_n = 1 + \frac{1}{2^2} + \frac{1}{3^2} + \frac{1}{4^2} + \dots + \frac{1}{n^2}$$

$$< 1 + \frac{1}{1 \times 2} + \frac{1}{2 \times 3} + \frac{1}{3 \times 4} + \dots + \frac{1}{(n-1) \times n} = 1 + \left(1 - \frac{1}{n}\right) < 2.$$

This shows that the sequence of partial sums is bounded above. It is increasing, because the terms of the series are positive, so to get from s_n to s_{n+1} we are always adding a positive number. Therefore the sequence (s_n) has a limit, i.e. the series converges. In fact, the sum of this series is $\pi^2/6$, but this is not easy to prove.

The technique used in this example can be generalized as follows.

● Proposition 2—The comparison test

Suppose that Σa_i is a convergent series of non-negative terms, with sum S, and that Σb_i is a series having the property that $0 \le b_i \le a_i$ for all $i \in \mathbb{N}$. Then Σb_i converges, with sum T, where $T \le S$.   ●

PROOF
Let (s_n) denote the sequence of partial sums for the series Σa_i and let (t_n) denote the sequence of partial sums for the series Σb_i. Because the terms of both series are non-negative, both sequences of partial sums are increasing. Because Σa_i converges, $\lim(s_n) = S$ and moreover $s_n \le S$ for all n. Since $b_i \le a_i$ for all i it follows that $t_n \le s_n \le S$ for all n. So (t_n) is a sequence which is increasing and bounded above (by S). It therefore converges to a limit T which cannot be greater than S.   ●

TUTORIAL PROBLEM 3

Show that in the comparison test the conclusion is true under the apparently weaker hypothesis that $0 \le b_i \le a_i$ for all $i \ge N$ for some positive integer N, rather than the inequality being satisfied for all values of i. This modification is typical of many situations involving limiting behaviour, where changing a finite number of terms of a series makes no difference to convergence.

The contrapositive of the comparison test is useful as a test for divergence. Under the same hypothesis $0 \le b_i \le a_i$ for all i, if Σb_i diverges then the sequence (t_n) tends to infinity. The sequence (s_n) therefore tends to infinity, and so Σa_i diverges.

● Proposition 3—The algebra of series

Suppose that Σa_n and Σb_n are convergent series with sums S and T respectively. Then

(i) the series $\Sigma(a_n + b_n)$ converges with sum $S + T$,

(ii) the series $\Sigma k a_n$ converges with sum kS, where k is a constant.   ●

PROOF

These results follow by applying the algebra of sequences (Proposition 1 of Chapter 7) to the sequences of partial sums.

TUTORIAL PROBLEM 4

We include here some results relating to Proposition 2.

(i) Show that if Σa_n diverges then $\Sigma k a_n$ diverges, where k is a non-zero constant.

(ii) Proposition 1(v) of Chapter 7 concerns products of sequences. Products of series are not so straightforward, and the fact that Σa_n converges to A and Σb_n converges to B does not imply that $\Sigma a_n b_n$ converges to AB; indeed it may not converge at all. Using the result of Proposition 6 below will tell us that if $a_n = b_n = (-1)^n/\sqrt{n}$ then Σa_n and Σb_n both converge. Use the result of Example 4 to show that $\Sigma a_n b_n$ diverges in this case.

Use the comparison test and the result of Proposition 1 to show that if Σa_n and Σb_n both converge, and if a_n and b_n are non-negative for all n, then $\Sigma a_n b_n$ converges.

Example 6

Discuss convergence of the series whose nth terms are given by the following formulae, using the comparison test.

(i) $a_n = \dfrac{n^3 + 4n + 3}{\sqrt{(n^{10} + n^7)}}$, (ii) $b_n = \dfrac{n^3 + 4n + 3}{\sqrt{(n^8 + 3n^7)}}$, (iii) $c_n = \dfrac{n^2 - 2n}{\sqrt{(n^6 + 3n^2 + 5)}}$.

As with Example 6(i) of Chapter 7, we use the strategy here of looking for the parts of the expression tending to infinity most rapidly, in order to try to gain an intuitive idea of how the series will behave before attempting a precise application of the comparison test.

(i) When n is very large the numerator is dominated by the n^3 component. The denominator is governed by $\sqrt{n^{10}} = n^5$. So, effectively, we have something which behaves like n^3/n^5, i.e. like $1/n^2$, which suggests convergence, by reference to Example 5. In applying the comparison test we therefore need to find an expression behaving like $1/n^2$ which is larger than a_n. We need to take care over inequalities involving algebraic fractions (Exercise 5 of §4.1 is relevant here). These considerations lead us to the following algebra.

$$0 < a_n = \frac{n^3 + 4n + 3}{\sqrt{(n^{10} + n^7)}} < \frac{n^3 + 4n^3 + 3n^3}{\sqrt{(n^{10})}} = \frac{8}{n^2}.$$

Example 5 and Proposition 3(ii) tell us that $\Sigma 8/n^2$ converges, and so Σa_n converges by comparison.

(ii) This time the numerator again behaves like n^3, but the denominator is governed by $\sqrt{n^8} = n^4$. So b_n is behaving like $1/n$, suggesting divergence, by

Example 4. This time we need to find an expression smaller than b_n behaving like $1/n$, in order to prove divergence.

$$b_n = \frac{n^3 + 4n + 3}{\sqrt{(n^8 + 3n^7)}} > \frac{n^3}{\sqrt{(n^8 + 3n^8)}} = \frac{1}{2n}.$$

Example 4 and Tutorial Problem 4(i) tell us that $\Sigma(1/2n)$ diverges, and so Σb_n diverges, by comparison.

(iii)   The relative algebraic degrees of the numerator and denominator again suggest behaviour like $1/n$, indicating divergence. This time we need to be a little more careful with the inequalities. We cannot simply omit all the components apart from n^2 from the denominator, for this will yield an inequality the wrong way round. We do want a comparison of the numerator with something of order n^2, and we can, in fact, say that

$$n^2 - 2n > n^2/2 \quad \text{provided} \quad n^2 > 4n, \quad \text{i.e.} \quad n > 2.$$

So for $n > 2$ we have

$$c_n = \frac{n^2 - 2n}{\sqrt{(n^6 + 3n^2 + 5)}} > \frac{n^2/2}{\sqrt{(n^6 + 3n^6 + 5n^6)}} = \frac{1}{6n}.$$

As with (ii), Example 4 and Tutorial Problem 4(i) tell us that $\Sigma(1/6n)$ diverges, and so Σc_n diverges, by comparison.

Example 7

Prove by comparison that $\Sigma(1/n^p)$ is convergent if $p > 1$ and divergent if $p \leq 1$.

This is a generalization of Examples 4 and 5. We will deal with divergence first. For $p \leq 1$, $1/n^p \geq 1/n$, and so using the result of Example 4 tells us that $\Sigma(1/n^p)$ is divergent, by comparison.

Now suppose that $p > 1$. We group the terms for comparison in a similar way to the method used in Example 4. We write out the first few terms explicitly to make the reasoning clear.

$$1 + \frac{1}{2^p} + \frac{1}{3^p} + \frac{1}{4^p} + \frac{1}{5^p} + \frac{1}{6^p} + \frac{1}{7^p} + \frac{1}{8^p} + \frac{1}{9^p} + \frac{1}{10^p} + \ldots + \frac{1}{16^p} + \ldots$$

$$< 1 + \frac{1}{2^p} + \frac{1}{2^p} + \frac{1}{4^p} + \frac{1}{4^p} + \frac{1}{4^p} + \frac{1}{4^p} + \frac{1}{8^p} + \frac{1}{8^p} + \frac{1}{8^p} + \ldots + \frac{1}{8^p} + \ldots$$

$$= 1 + \frac{2}{2^p} + \frac{4}{4^p} + \frac{8}{8^p} + \ldots = 1 + \frac{1}{2^{p-1}} + \frac{1}{(2^{p-1})^2} + \frac{1}{(2^{p-1})^3} + \ldots = \frac{1}{1 - \left(\frac{1}{2}\right)^{p-1}}$$

by summing the geometric series whose common ratio is $1/2^{p-1}$, which is less than 1 because $p - 1 > 0$. Therefore, by comparison, the original series converges.

The series $\Sigma(1/n^p)$ defines a function of the variable p for all real values of $p > 1$. This function has been studied extensively, in particular at the turn of the century in connection with the distribution of prime numbers. It is known as the Riemann Zeta-function, denoted by $\zeta(p)$, and named after the German mathematician

Riemann (1826–1866). In fact the series can be generalized to allow p to be any complex number with real part greater than 1.

In Example 8 we used a geometric series for comparison. This can also be generalized, as follows.

● *Proposition 4—d'Alembert's ratio test*

We shall give this in two forms.

(i) If the series Σa_n has all its terms positive, and if there is a number $k < 1$, independent of n, such that the ratio $a_{n+1}/a_n \le k$ for all values of n exceeding some number N, then the series converges.

(ii) If the series Σa_n has all its terms positive, and if the ratio a_{n+1}/a_n has a limit $l < 1$, then the series converges.   ●

PROOF

We shall use the comparison test as modified by the result of Tutorial Problem 3. For $n > N$ we can write

$$a_n = \frac{a_n}{a_{n-1}} \cdot \frac{a_{n-1}}{a_{n-2}} \cdot \frac{a_{n-2}}{a_{n-3}} \cdots \frac{a_{N+1}}{a_N} \cdot a_N \le k^{n-N} a_N = k^n a_N k^{-N}.$$

The right-hand side is the nth term of a geometric series with common ratio $k < 1$, which therefore converges. So Σa_n converges by comparison.

(ii)   Let $\epsilon = (1 - l)/2 > 0$. Let $k = l + \epsilon = (1 + l)/2 < 1$. Because $\lim (a_{n+1}/a_n) = l$, it follows that there is a number N so that for all $n > N$, $|a_n - l| < \epsilon$, i.e. $a_{n+1}/a_n < k$. So the hypothesis of (i) is satisfied and hence the series converges.   ●

It is important to note that we must have k and l strictly less than 1. For example, if we let $a_n = 1/n$ then $a_{n+1}/a_n < 1$ for all n, and also $\lim(a_{n+1}/a_n) = 1$. The series Σa_n diverges, as we already know from Example 4. However, if we take $a_n = 1/n^2$ the limit of the ratio of successive terms is again one, but the series converges.

We conclude that if the limit of the ratio is equal to 1 this gives us no information about convergence.

It is natural to ask what happens if the limit of the ratio is greater than 1. In this case we conclude that the terms are eventually increasing. They do not therefore tend to zero and so the series diverges by the contrapositive of Proposition 1.

● *Example 8.*

Show that the following series converge, using the ratio test.

(i) $\displaystyle\sum \frac{(n!)^2}{(2n)!} \cdot \frac{3^{2n}}{2^{2n}}$,   (ii) $\displaystyle\sum \frac{n+1}{n!}$,   (iii) $\displaystyle\sum \frac{n!}{10^n}$.

(i)   We calculate the ratio of successive terms to obtain

$$\frac{a_{n+1}}{a_n} = \frac{(n+1)!(n+1)!\,3^{2n+2}}{(2n+2)!\,2^{2n+2}} \cdot \frac{(2n)!\,2^{2n}}{n!n!\,3^{2n}} = \frac{(n+1)(n+1)3^2}{(2n+2)(2n+1)2^2} \rightarrow \frac{9}{16} < 1.$$

Therefore the series converges, by the second form of the ratio test.

(ii)   We again calculate the ratio of successive terms to obtain

$$\frac{a_{n+1}}{a_n} = \frac{n+2}{(n+1)!} \cdot \frac{n!}{n+1} = \frac{n+2}{(n+1)^2} \rightarrow 0 < 1.$$

Therefore the series converges. Notice that in the statement of the test the only requirement is that $l < 1$, so that $l = 0$ satisfies the conditions.

(iii)   The ratio of successive terms in this case gives

$$\frac{a_{n+1}}{a_n} = \frac{(n+1)!}{10^{n+1}} \cdot \frac{10^n}{n!} = \frac{n+1}{10}.$$

This ratio tends to infinity and so the series diverges. In fact it is enough to note that the ratio is bigger than 1 for $n > 9$, and so the equivalent for divergence of form (i) of the ratio test applies. It is interesting to note that in fact the ratio is less than 1 for the first nine cases, so that the terms of the series decrease for a while, but then increase, and in fact the terms themselves tend to infinity, so that we could also establish divergence by using the contrapositive of Proposition 1.

● Proposition 5—Cauchy's root test

Like the ratio test we shall give this in two forms.

(i) If the series Σa_n has all its terms positive, and if there is a number $k < 1$, independent of n, such that $\sqrt[n]{a_n} \le k$ for all values of n exceeding some number N, then the series converges.

(ii) If the series Σa_n has all its terms positive, and if $\sqrt[n]{a_n}$ has a limit $l < 1$, then the series converges.   ●

PROOF

We shall prove part (i) only. Part (ii) follows from part (i) in exactly the same way as the corresponding deduction for the ratio test, and we leave the details to the reader. Like the ratio test this result involves comparison with a geometric series, since $\sqrt[n]{a_n} \le k$ is equivalent to $a_n \le k^n$, and since $k < 1$ the geometric series Σk^n converges, implying convergence of Σa_n by comparison.   ●

As with d'Alembert's ratio test, it is important that k and l are strictly less than 1. The same examples we used in that connection demonstrate that Cauchy's root test is also inconclusive if k or l is equal to 1. If $l > 1$ or if $\sqrt[n]{a_n} \ge 1$ for all sufficiently large n then $a_n \nrightarrow 0$ and so the series diverges.

So far the tests in this section have involved series all of whose terms are non-negative. We finish with a special test for series whose terms alternate in sign. Such series occur sometimes in approximations to functions.

● *Proposition 6—Leibniz's alternating test*

If (a_n) is a decreasing sequence of positive real numbers with limit zero then the series $\Sigma(-1)^n a_n$ converges.   ●

PROOF

We consider first the sequence of even partial sums, bracketed in two ways.

$$s_{2n} = a_1 - a_2 + a_3 - a_4 + \ldots + a_{2n-1} - a_{2n}$$
$$= (a_1 - a_2) + (a_3 - a_4) + \ldots + (a_{2n-1} - a_{2n})$$
$$= a_1 - (a_2 - a_3) - (a_4 - a_5) - \ldots - (a_{2n-2} - a_{2n-1}) - a_{2n}.$$

We use the hypothesis that $a_1 \geq a_2 \geq a_3 \geq \ldots$. The first bracketing then shows that (s_{2n}) is an increasing sequence. The second bracketing shows that $s_{2n} \leq a_1$ for all n. So the even partial sums form an increasing sequence which is bounded above, and it therefore has a limit. We now observe that $s_{2n+1} = s_{2n} + a_{2n+1}$, and so the fact that $\lim(a_n) = 0$ tells us that the sequence of odd partial sums has the same limit as the sequence of even partial sums. Therefore, the entire sequence of partial sums has a limit, proving that the series is convergent.   ●

● *Example 9*

Prove that the series $1 - \frac{1}{2} + \frac{1}{3} - \frac{1}{4} + \frac{1}{5} - \frac{1}{6} + \ldots$ is convergent.

This is a straightforward application of Leibniz's test. The general term of the series is $(-1)^n/n$. The sequence $(1/n)$ is decreasing with limit zero, so that the series satisfies all the conditions for the test and is therefore convergent. (In fact its sum is $\ln(2)$.) Note that without the alternating signs, in other words with all the terms positive, we have $\Sigma 1/n$, which diverges.

The condition that (a_n) should be a decreasing sequence used in Leibniz's test cannot be dropped. If we consider the series

$$1 - \frac{1}{2} + \frac{1}{3} - \frac{1}{4} + \frac{1}{5} - \frac{1}{8} + \frac{1}{7} - \frac{1}{16} + \ldots,$$

we find that the positive terms form a divergent series, whereas the negative terms form a convergent geometric series with commmon ratio $\frac{1}{2}$. The complete series therefore diverges, for the sequence of partial sums will tend to infinity.

There are a number of other tests for convergence which have been developed. They are rather specialized, and can be found in older books (see §10.5 for references).

EXERCISES 8.2

1. Determine the convergence or otherwise of the series Σa_n, where a_n is defined by the following formulae.

(i) $\dfrac{1}{2n+1}$,   (ii) $\dfrac{1}{n^2+1}$,   (iii) $\dfrac{1}{\sqrt{(n(n+1))}}$,   (iv) $\sqrt{n+1}-\sqrt{n}$,

(v) $\dfrac{n^2-2n+3}{n^4-4n+1}$,   (vi) $\dfrac{\sqrt{n+1}-\sqrt{n}}{n+1}$,   (vii) $\dfrac{5^n}{3^{n-1}2^{n+1}}$,

(viii) $\dfrac{n+3^n}{n^6+2^n}$,   (ix) $\dfrac{\sqrt{n}}{\ln(n)}$,   (x) $\dfrac{(-1)^n}{\ln(n)}$.

2. Use the ratio test to determine whether the following series converge. In each case write down a formula for the general term a_n following the numerical patterns given.

(i) $\dfrac{2}{3}+2\left(\dfrac{2}{3}\right)^2+3\left(\dfrac{2}{3}\right)^3+4\left(\dfrac{2}{3}\right)^4+\cdots$

(ii) $\dfrac{5}{2}+\dfrac{5^2}{2.2^2}+\dfrac{5^3}{3.2^3}+\dfrac{5^4}{4.2^4}+\cdots$

(iii) $\dfrac{1}{3}+\dfrac{1.2}{3.5}+\dfrac{1.2.3}{3.5.7}+\dfrac{1.2.3.4}{3.5.7.9}+\cdots$

(iv) $7+\dfrac{7^2}{2!}+\dfrac{7^3}{3!}+\dfrac{7^4}{4!}+\cdots$

(v) $\dfrac{1}{9}+\dfrac{2!}{9^2}+\dfrac{3!}{9^3}+\dfrac{4!}{9^4}+\cdots$

(vi) $\dfrac{1}{2}+\dfrac{2.2!}{5}+\dfrac{2^2.3!}{10}+\dfrac{2^3.4!}{17}+\dfrac{2^4.5!}{26}+\dfrac{2^5.6!}{37}+\cdots$.

3. Use the algebra of series (Proposition 3) to express the series

$$\sum_{r=1}^{\infty}\frac{(r+1)^2}{r!}$$

in terms of the exponential series, to show that its sum is $5e-1$.

8.3 Series and Integrals

For some regularly behaved series it is possible to make comparisons with integrals, in particular if the series is of the form $\Sigma f(n)$, where $f(x)$ is a positive decreasing function of the real variable x. The situation is illustrated in Fig 8.2, where we compare the areas of rectangles with that of the shaded region.

The rectangles we use have base AB, which has length 1, and heights equal to the value of the function at the appropriate points. The shaded region has its area given by an integral, so we have

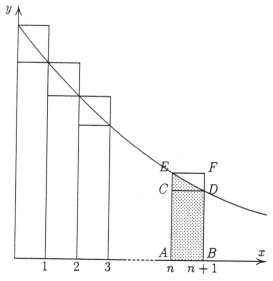

Fig 8.2 Comparing series and integrals through areas.

area (rectangle $ACDB$) $\leq$ shaded area $\leq$ area (rectangle $AEFB$),

$$f(n+1) \leq \int_n^{n+1} f(x)\,dx \leq f(n).$$

Copying the right-hand inequality, and replacing n by $n-1$ in the left-hand inequality enables us to deduce that

$$\int_n^{n+1} f(x)\,dx \leq f(n) \leq \int_{n-1}^n f(x)\,dx.$$

If we sum each of these chains of inequalities for n from 1 to k we obtain

$$\sum_{n=2}^{k+1} f(n) \leq \int_1^{k+1} f(x)\,dx \leq \sum_{n=1}^k f(n), \tag{I}$$

$$\int_1^{k+1} f(x)\,dx \leq \sum_{n=1}^k f(n) \leq \int_0^k f(x)\,dx. \tag{II}$$

Notice first that if we imagine that we can perform the indefinite integration and then substitute the limits, the answer will contain the integer variable k. In other words, the integrals themselves form a sequence of real numbers, and we can discuss convergence of such sequences. We can conclude a number of things from the inequalities, using the fact that both the sums and the integrals increase as k increases, because f is a positive function. These deductions can be summarized in Proposition 7.

● *Proposition 7—The integral test*

Suppose that $f(x)$ is a positive decreasing function of the real variable x. The infinite series $\Sigma f(n)$ converges if and only if the sequence of integrals $\displaystyle\int_1^k f(x)\,dx$ converges

as k tends to infinity. ●

PROOF

(a) Suppose that $\displaystyle\lim_k\left(\int_0^k f(x)\,dx\right) = l$, where we have used the notation $\displaystyle\lim_k$ to

draw particular attention to the fact that k is the integer variable in the sequence
defined by the integral. Then from the right-hand inequality of (II) we deduce that

$$\sum_{n=1}^k f(n) < l \text{ for all } k,$$ so the associated infinite series converges, because the

sequence of partial sums is increasing and bounded above.

(b) Similar reasoning to that of (a) applied to the right-hand side of (I) tells us that
if the infinite series converges then the integral also tends to a limit as k tends to
infinity.

(c) We can also conclude, from the left-hand side of (I), that if the infinite series
diverges then the sequence of integrals also diverges.

(d) Finally, we deduce from the left-hand side of (II) that if the sequence of
integrals diverges so does the infinite series. ●

We can make a further deduction from the left-hand side of (II).

$$T_k = \sum_{n=1}^k f(n) - \int_1^k f(x)\,dx \geq \int_k^{k+1} f(x)\,dx > 0.$$

So the sequence (T_k) is bounded below. Moreover

$$T_{k+1} - T_k = f(k+1) - \int_k^{k+1} f(x)\,dx < 0$$

from the original comparison of areas. So the sequence (T_k) is decreasing and
bounded below, and therefore has a limit. This shows that

$$\sum_{n=1}^k f(n) - \int_1^k f(x)\,dx$$

has a limit as k tends to infinity whether the series converges or diverges. The next
example is an application of this result.

● *Example 10*

Show that $\displaystyle\sum_{n=1}^k \frac{1}{n} - \ln(k)$ has a limit.

We apply the above result with $f(x) = 1/x$, so that

$$\sum_{n=1}^{k} f(n) - \int_{1}^{k} f(x)\,dx = \sum_{n=1}^{k} \frac{1}{n} - \ln(k).$$

The limit in question is, in fact, a well-known constant, known as Euler's constant, usually denoted by the Greek letter γ, and having approximate value 0.5772. It is not known whether Euler's constant is a rational number.

We can use the techniques developed here to investigate the rate of convergence, i.e. how many terms of a series we need to add in order to approximate to the sum with a specified degree of accuracy. We shall consider this through examples.

Example 11

How accurate an estimate do we get if we add the first 10 terms of the series $\Sigma 1/n^2$? How many terms do we need to achieve an accuracy of at least 10^{-4}?

In the same way as we deduced (II) from the preceding inequalities we have

$$\int_{m+1}^{k+1} \frac{1}{x^2}\,dx \le \sum_{n=m+1}^{k} \frac{1}{n^2} \le \int_{m}^{k} \frac{1}{x^2}\,dx.$$

Evaluating the integrals then gives

$$\left(\frac{1}{m+1} - \frac{1}{k+1}\right) \le \sum_{n=m+1}^{k} \frac{1}{n^2} \le \left(\frac{1}{m} - \frac{1}{k}\right).$$

Taking the limit as k tends to infinity gives

$$\frac{1}{m+1} \le \sum_{n=m+1}^{\infty} \frac{1}{n^2} \le \frac{1}{m}.$$

Now the sum here is the difference between the sum of the complete infinite series and the result of adding the first m terms, and so it gives a measure of the error involved in adding just these first m terms. We shall denote this by E_m. If we take $m = 10$ this shows that

$$\frac{1}{11} \le E_{10} \le \frac{1}{10}.$$

So adding the first 10 terms gives an error within these limits. If, to this total, we add $1/11$ the error would be between 0 and $1/10 - 1/11 \approx 0.009$. Performing this calculation gives an estimate of 1.6407. We have already quoted the sum of the infinite series as $\pi^2/6$ which is approximately 1.6449. This confirms the result of the error analysis. To obtain a partial sum with an accuracy of 10^{-4} we would need to find m such that $E_m \le 10^{-4}$, i.e. $1/m < 10^{-4}$, giving $m > 10000$, so that convergence is fairly slow.

Example 12

Apply the integral test to investigate the convergence of the series

(i) $\displaystyle\sum_{n=1}^{\infty}\frac{1}{n^p}$, (ii) $\displaystyle\sum_{n=2}^{\infty}\frac{1}{n(\ln(n))^p}$,

where $p > 0$. Investigate the rate of convergence in case (ii) when $p = 2$.

(i) We dealt with this situation in Example 7, and so this provides some confirmation of the integral test in action. We let $f(x) = 1/x^p$. This is a positive decreasing function when p is positive. Calculating the integral gives, for $p \neq 1$,

$$\int_1^k \frac{1}{x^p}\,dx = \left[\frac{-1}{(p-1)x^{p-1}}\right]_{x=1}^{x=k} = \frac{1}{p-1}\left(1 - \frac{1}{k^{p-1}}\right).$$

This has a limit $(1/(p-1))$ if $p > 1$, but diverges if $p < 1$. For $p = 1$ the integral gives $\ln(k)$ which also diverges. So the integral test tells us that the series converges for $p > 1$ and diverges for $p \leq 1$.

(ii) Notice that in this case the summation begins with $n = 2$ simply because the function is undefined when $n = 1$. Again working out the integral gives

$$\int_2^k \frac{1}{x(\ln(x))^p}\,dx = \left[\frac{-1}{(p-1)(\ln(x))^{p-1}}\right]_{x=2}^{x=k} = \frac{1}{p-1}\left(\frac{1}{(\ln(2))^{p-1}} - \frac{1}{(\ln(k))^{p-1}}\right).$$

This has a limit $(\ln(2))^{-(p-1)}/(p-1)$ if $p > 1$ but diverges if $p < 1$. For $p = 1$ the integral becomes

$$\int_2^k \frac{1}{x\ln(x)}\,dx = \ln(\ln(k)) - \ln(\ln(2)) \to \infty.$$

So the integral test tells us that the series converges for $p > 1$ and diverges for $p \leq 1$.

To investigate the rate of convergence when $p = 2$ we use the method of Example 11. So

$$\int_{m+1}^{k+1}\frac{1}{x(\ln(x))^2}\,dx \leq \sum_{n=m+1}^{k}\frac{1}{n(\ln(n))^2} \leq \int_m^k \frac{1}{x(\ln(x))^2}\,dx.$$

$$\frac{1}{\ln(m+1)} - \frac{1}{\ln(k+1)} \leq \sum_{n=m+1}^{k}\frac{1}{n(\ln(n))^2} \leq \frac{1}{\ln(m)} - \frac{1}{\ln(k)}.$$

Taking the limit gives

$$\frac{1}{\ln(m+1)} \leq \sum_{n=m+1}^{\infty}\frac{1}{n(\ln(n))^2} \leq \frac{1}{\ln(m)}.$$

So to obtain a partial sum which is within 10^{-4} of the infinite sum we need $\ln(m) > 10^4$, i.e. $m > e^{(10^4)} \approx 10^{4343}$. Indeed, to obtain a partial sum within 10^{-1} of the infinite sum we need $m > e^{10} \approx 22026$. So convergence is extremely slow.

Having seen some slowly convergent series we can investigate the same question for a geometric series. In this case the difference between the partial sum s_n and the infinite sum is $r^{n+1}/(1-r)$. So to obtain a partial sum within 10^{-4} of the infinite sum we need to solve the inequality $r^{n+1}/(1-r) < 10^{-4}$. We leave the detailed solution of this inequality to the reader. When $r = 0.9$ it gives $n \geq 109$, and when $r = 0.5$ we need $n \geq 14$, so convergence is quite quick.

EXERCISES 8.3

1. Use the integral test to find the values of p for which the following series converges

$$\sum_{r=3}^{\infty} \frac{1}{r \ln(r)(\ln(\ln(r)))^p}.$$

2. Estimate the number of terms needed to find the sums of the following convergent series correct to within 10^{-4}.

   (i) $\displaystyle\sum_{n=1}^{\infty} n^{-5/3}$,   (ii) $\displaystyle\sum_{n=2}^{\infty} \frac{1}{n(\ln(n))^4}$.

8.4 Complex Series and Absolute Convergence

We remarked after Definition 1 that the notion of an infinite series is applicable when the terms are complex numbers. We discussed sequences of complex numbers in §7.6, and proved there that a complex sequence converges if and only if the sequences of real and imaginary parts both converge.

TUTORIAL PROBLEM 5

Show that if $z_n = x_n + y_n i$, then the complex series Σz_n converges if and only if the real series Σx_n and Σy_n both converge, using the result of Proposition 5 in Chapter 7.

This means that we can test for convergence by trying to apply the tests of the previous section to the series of real and imaginary parts separately. A further aspect of convergence, which is important in relation to both real and complex series is that of absolute convergence, defined as follows.

● Definition 2

An infinite series Σz_n of complex or real terms is said to be absolutely convergent if the series (of non-negative real numbers) $\Sigma |z_n|$ converges. A series which is convergent but not absolutely convergent is said to be conditionally convergent.   ●

In testing for absolute convergence we are concerned with $\Sigma |z_n|$, which is a series with non-negative real terms. The various tests developed in the previous section are therefore directly applicable. Referring back to Example 9 shows a series, $\Sigma(-1)^n/n$, which is convergent but not absolutely convergent. The next proposition demonstrates that the set of absolutely convergent series is a subset of the set of convergent series. Before proving it we shall analyse real series in relation to the positive and negative terms separately.

Let Σa_n be a series of real terms, and let

$$a_n^+ = \begin{cases} a_n, & \text{if } a_n > 0; \\ 0, & \text{if } a_n \le 0; \end{cases} \qquad a_n^- = \begin{cases} -a_n, & \text{if } a_n < 0; \\ 0, & \text{if } a_n \ge 0. \end{cases}$$

TUTORIAL PROBLEM 6

Explain why $a_n = a_n^+ - a_n^-$ and why $|a_n| = a_n^+ + a_n^-$, for all n.

● Proposition 8

If the series Σz_n is absolutely convergent then it is convergent. ●

PROOF

We decompose the terms into their real and imaginary parts, so let $z_n = x_n + y_n i$. Now for all n, $|x_n| \le |z_n|$ and $|y_n| \le |z_n|$. So by comparison $\Sigma |x_n|$ and $\Sigma |y_n|$ are both convergent. Furthermore, $0 \le x_n^+ \le |x_n|, 0 \le x_n^- \le |x_n|, 0 \le y_n^+ \le |y_n|$ and $0 \le y_n^- \le |y_n|$, and so by comparison the series Σx_n^+, Σx_n^-, Σy_n^+ and Σy_n^- all converge. The first result from Tutorial Problem 6, and the algebra of series (Proposition 3) now tell us that Σx_n and Σy_n both converge, and so finally Σz_n converges. ●

The importance of this idea lies in the fact that operations which are valid for finite arithmetic addition are generally valid for absolutely convergent infinite series. So, given such a series we can bracket the terms however we like, and add within the brackets first, and the new series will have the same sum as the old. We can add the terms in a totally different order and still have the same sum. Proving such results in general is outside the scope of this book, but we shall give one example of a conditionally convergent series where changing the order of summation actually changes the sum. We consider the series

$$1 - \frac{1}{2} + \frac{1}{3} - \frac{1}{4} + \frac{1}{5} - \frac{1}{6} + \dots = \ln(2),$$

where the terms are alternately positive and negative. We shall use the result of Example 10, so if we let $\gamma_n = 1 + (1/2) + (1/3) + \dots + (1/n) - \ln(n)$ then $\lim(\gamma_n) = \gamma$ (Euler's constant). Now let us add the series by taking a positive term followed by two negative terms, to give

$$1 - \frac{1}{2} - \frac{1}{4} + \frac{1}{3} - \frac{1}{6} - \frac{1}{8} + \frac{1}{5} - \dots$$

All the terms of the original series are present, indeed it would be possible to give a formula to indicate the position of each in the rearrangement. We consider the partial sum consisting of $3n$ terms,

$$
\begin{aligned}
t_{3n} &= 1 - \frac{1}{2} - \frac{1}{4} + \frac{1}{3} - \frac{1}{6} - \frac{1}{8} \cdots + \frac{1}{2n-1} - \frac{1}{4n-2} - \frac{1}{4n} \\
&= \left(1 + \frac{1}{3} + \frac{1}{5} + \ldots + \frac{1}{2n-1}\right) - \left(\frac{1}{2} + \frac{1}{4} + \ldots + \frac{1}{4n}\right) \\
&= \left(1 + \frac{1}{2} + \frac{1}{3} + \ldots + \frac{1}{2n}\right) - \left(\frac{1}{2} + \frac{1}{4} + \ldots + \frac{1}{2n}\right) - \left(\frac{1}{2} + \frac{1}{4} + \ldots + \frac{1}{4n}\right) \\
&= (\gamma_{2n} + \ln(2n)) - (\gamma_n + \ln(n))/2 - (\gamma_{2n} + \ln(2n))/2 \\
&= (\gamma_{2n} - \gamma_n)/2 + \ln(2)/2 \to \ln(2)/2.
\end{aligned}
$$

Considering the remaining partial sums gives

$$
t_{3n+1} = t_{3n} + \frac{1}{2n+1} \to \ln(2)/2; \quad t_{3n+2} = t_{3n} + \frac{1}{2n+1} - \frac{1}{4n+2} \to \ln(2)/2.
$$

This shows that the sum of the rearranged series is half that of the original one. This is a startling result. The idea that we can add a collection of numbers and, by changing the order of addition, change the sum is totally different from finite arithmetic, and serves to illustrate that the limiting processes involved in dealing with infinite series need careful analysis whenever they are used. We can go beyond this example, for a result known as Riemann's Rearrangement Theorem says that a conditionally convergent series can be rearranged so that its sum is equal to any specified number. It can also be rearranged so as to diverge to infinity, or to negative infinity. It can also be rearranged so that the sequence of partial sums oscillates between any two specified bounds. In other words, we can make it do almost anything we might think of! None of this happens with absolutely convergent series—they behave very respectably. One of the main contexts where absolute convergence is encountered is that of power series, which we discuss in the next section.

8.5 Power Series

If you use a function button on your calculator, for example the 'sin' key, it is natural to ask how it works. The calculator will not have a table of the sine function built in, and indeed basically all it can do is addition. This of course extends to multiplication as repeated addition. So to deal with the sine function an approximation has to be computed using the basic operations of arithmetic. The idea of such approximations was investigated long before the advent of calculators, indeed centuries ago, and so the theory of such approximations is well developed. The usefulness of these approximations lies in the capability they have for including a real variable, and to be valid over some interval of real numbers, rather than relating to just one number, like the exponential series discussed at the end of §8.1. In fact, much of the theory is valid for a complex variable, which we shall use.

• Definition 3

A power series is an infinite series of the form $\Sigma a_n z^n$, where z is a complex variable and (a_n) is a sequence of complex numbers, referred to as the coefficients. The summation conventionally begins with $n = 0$, since $z^0 = 1$ allows a constant term a_0. In the power series representation of the elementary functions, the coefficients are invariably real numbers, and we shall use real coefficients in the examples. ●

Example 13

Find the values of z for which the power series $\displaystyle\sum_{n=0}^{\infty} \frac{(n+1)^2}{2^n} z^n$ converges absolutely.

We use the ratio test. The ratio of the absolute value of successive terms is

$$\frac{(n+2)^2|z|^{n+1}}{2^{n+1}} \cdot \frac{2^n}{(n+1)^2|z|^n} = \left(\frac{n+2}{n+1}\right)^2 \frac{|z|}{2} \to \frac{|z|}{2}.$$

Thus, the series converges absolutely if $|z|/2 < 1$, i.e. $|z| < 2$, and diverges if $|z|/2 > 1$, ie $|z| > 2$. If $|z| = 2$ the ratio test is inconclusive. However, in that case, the ratio is in fact greater than one for all n, so that the terms increase in magnitude and therefore will not tend to zero. We therefore have divergence if $|z| = 2$.

We notice in this example that the values of z for which we have convergence satisfy $|z| < 2$. This is the interior of a circle in the complex plane. In fact this happens almost universally, as we shall prove in the next two propositions.

• Proposition 9

If the power series $\Sigma a_n z^n$ converges for $z = C$ then the series converges absolutely for all values of z satisfying $|z| < |C|$. ●

PROOF

Since $\Sigma a_n C^n$ converges, $\lim(a_n C^n) = 0$. From the definition of a limit, taking $\epsilon = 1$, there is a positive integer M such that for all $n > M$, $|a_n C^n| < 1$. We now define a number K by

$$K = \max\{|a_0|, |a_1 C|, |a_2 C^2|, \ldots, |a_M C^M|, 1\}.$$

We then have $|a_n C^n| \le K$ for all n. Thus, for all n and for $|z| < C$

$$|a_n z^n| = |a_n C^n|\left|\frac{z}{C}\right|^n \le K\left|\frac{z}{C}\right|^n.$$

The right-hand expression is the nth term of a geometric series with common ratio $|z/C|$, less than 1, and therefore convergent. So $\Sigma a_n z^n$ is absolutely convergent, by comparison. ●

TUTORIAL PROBLEM 7

By considering the contrapositive of the last result show that if $\Sigma a_n D^n$ diverges then $\Sigma a_n z^n$ diverges for all values of z satisfying $|z| > |D|$.

• Proposition 10

Let $\Sigma a_n z^n$ be a power series. Then there are three possibilities.

(a) It converges absolutely for all z.

(b) It converges only for $z = 0$.

(c) There is a positive real number R such that the series converges absolutely if $|z| < R$ and diverges if $|z| > R$. The number R is called the radius of convergence of the power series. The circle $|z| = R$ is called the circle of convergence.   •

PROOF

(a)   The series $\Sigma z^n / n!$ converges absolutely for all z, as can be verified using the ratio test.

(b)   The series $\Sigma n! z^n$ diverges for all $z \neq 0$, as can be verified using the ratio test.

(c)   The proof of this result uses the axiom of completeness for the real number system discussed in Chapter 5. It uses the logic of the definition of least upper bound (Definition 4 of Chapter 5). We shall give the proof here as an important application of the concept of least upper bound. The most important thing to understand at this stage however is the significance of the result, and the subsequent examples will illustrate that.

Let S be the set of non-negative real numbers s having the property that $\Sigma a_n s^n$ is convergent. Having dealt with cases (a) and (b), S will not contain all positive real numbers, so that there is some number D not in S. Tutorial Problem 7 then tells us that no number greater than D can be in S, and so S is bounded above. Now $0 \in S$ (every power series converges when $z = 0$ as all the terms are zero, except perhaps the first). So S has a least upper bound, denoted by R.

It follows firstly that if $|z| > R$ then the power series diverges, for if not we would have numbers greater than R for which the series converged, and so R would not be an upper bound for S. Now let z_0 be any number satisfying $|z_0| < R$. We define the positive real number H by $H = (|z_0| + R)/2$. If the power series were divergent for $z = H$ then H would be an upper bound for S. But $H < R$ and so we conclude that the power series converges for $z = H$. Proposition 9 then tells us that the power series converges absolutely for $|z| < H$ and so in particular for $z = z_0$.   •

If we look back at Example 13 we can now see that we established that the power series in that example has radius of convergence 2.

Example 14

Find the radius of convergence of the following power series. Investigate what happens on the circle of convergence.

(i) $\displaystyle\sum_{n=1}^{\infty} \frac{z^n}{n^2}$,   (ii) $\displaystyle\sum_{n=0}^{\infty} 2^n z^n$,   (iii) $\displaystyle\sum_{n=1}^{\infty} \frac{z^n}{n}$,   (iv) $\displaystyle\sum_{n=0}^{\infty} (2 - (-1)^n) 2^n z^n$.

In each case we apply the ratio test for absolute convergence.

(i) $\left|\dfrac{z^{n+1}}{(n+1)^2}\right|\left|\dfrac{n^2}{z^n}\right| = \left|\dfrac{n^2}{(n+1)^2}\,z\right| \to |z|.$ So the series converges absolutely if $|z| < 1,$

and diverges if $|z| > 1.$ The radius of convergence is therefore equal to 1. For $|z| = 1$ the power series becomes $\Sigma(1/n^2)$ which converges. So the power series is absolutely convergent everwhere on the circle of convergence.

(ii)   The ratio test again gives $R = 1.$ This time however when $|z| = R$ we have $\Sigma|n^2 z^n| = \Sigma n^2,$ which diverges because the nth term does not tend to zero. So we have divergence everywhere on the circle $|z| = 1.$

(iii)   Once more the ratio test gives $R = 1.$ This time the behaviour on the circle of convergence varies. For $z = 1$ the series is $\Sigma(1/n)$ which diverges. For $z = -1$ the series becomes $\Sigma(-1)^n/n,$ which converges. A general analysis of the behaviour on the circle of convergence is somewhat involved, but readers may like to explore the behaviour when $z = i.$

(iv)   If we calculate the ratio of successive terms we obtain

$$\left|\frac{(2-(-1)^{n+1})2^{n+1}z^{n+1}}{(2-(-1)^n)2^n z^n}\right| = \left|\frac{(2-(-1)^{n+1})}{(2-(-1)^n)}\,2z\right|.$$

If n is even this reduces to $6|z|,$ while if n is odd it gives $2|z|/3.$ So the sequence of ratios does not have a limit, and we have to find another approach, using a comparison. The clue comes from realizing that the factor $2 - (-1)^n$ is bounded above, and bounded below by a positive number, and experience shows that replacing this by a constant makes no difference to the radius of convergence. So we use the inequality $2 - (-1)^n \le 3$ and consider the power series $\Sigma 3.2^n z^n.$ Applying the ratio test tells us that this series has radius of convergence equal to $\frac{1}{2},$ and so by comparison the original series converges absolutely for $|z| < \frac{1}{2}.$ So the radius of convergence for the original series is not less than $\frac{1}{2}.$ If we now substitute $z = \frac{1}{2}$ into the orginal power series we obtain $\Sigma(2 - (-1)^n).$ This series diverges, because the nth term does not tend to zero. To summarize, we have established that the power series converges absolutely if $|z| < \frac{1}{2},$ and diverges if $z = \frac{1}{2}.$ This demonstrates that the radius of convergence is equal to $\frac{1}{2}.$

The elementary functions can be defined by means of power series, and a finite number of terms used as a polynomial approximation as discussed at the beginning of this section. Proving these results is part of many calculus courses, under the heading of Taylor's Theorem. This is outside the scope of this book, but we record here these power series because they are so important.

$$e^z = 1 + z + \frac{z^2}{2!} + \frac{z^3}{3!} + \frac{z^4}{4!} + \cdots$$

$$\cos z = 1 - \frac{z^2}{2!} + \frac{z^4}{4!} - \frac{z^6}{6!} + \cdots$$

$$\sin z = z - \frac{z^3}{3!} + \frac{z^5}{5!} - \frac{z^7}{7!} + \cdots$$

These definitions are first encountered when z is replaced by x—a real variable (see §6.5). The definitions are equally applicable to complex variables however, and in the study of that topic we find that the exponential and trigonometric functions are much more closely related than the real variable situation would suggest.

EXERCISES 8.5

1. Show that the power series for the exponential, sine and cosine functions are absolutely convergent for all z.

2. Find the radius of convergence R of the power series $\Sigma a_n z^n$ where a_n is given by the formulae below. Note that in some cases the formula is undefined for $n = 0$. In those cases let $a_0 = 1$ for completeness sake, since the value of a_0 makes no difference to the radius of convergence. In each case determine whether the power series is convergent when $z = R$ and $z = -R$ (the two points where the circle of convergence intersects the real axis).

$$\text{(i)} \quad a_n = n^3, \quad \text{(ii)} \quad a_n = \frac{2^n}{n!}, \quad \text{(iii)} \quad a_n = \frac{2^n}{n^2},$$

$$\text{(iv)} \quad a_n = \frac{n^3}{2^n}, \quad \text{(v)} \quad a_n = \sum_{m=1}^{n} \frac{1}{m}, \quad \text{(vi)} \quad a_n = \frac{(-1)^{n+1}}{n},$$

$$\text{(vii)} \quad a_n = 1 + (-1)^n + \frac{1}{n}, \quad \text{(viii)} \quad a_n = \frac{(-1)^n}{n^2}, \quad \text{(ix)} \quad a_n = \left(1 + \frac{1}{n}\right)^n.$$

3. Show that the radius of convergence of the power series $\Sigma(n + 1)z^n$ is equal to 1. Show that inside the circle of convergence its sum is $(1 - z)^{-2}$.

Summary

The study of infinite series is motivated partly by the desire to extend the processes of arithmetic so as to give meaning to the notion of adding an infinite set of numbers. This has been done through the theory of limits of sequences explored in Chapter 7. A major part of the chapter has involved developing tests for convergence to deal with situations in which it is not possible to find explicit algebraic formulae for sums.

We emphasized the important fact that if we have a series for which we know that the nth term tends to zero, this does not tell us whether the series converges or diverges. Examples 4 and 5 were given in illustration of this.

The basic technique involved in the tests was that of comparison, whereby the convergence or otherwise of a series with algebraically complicated terms can be compared with one whose terms are simpler. Two specific tests, attributed to d'Alembert and Cauchy, were derived from this, principally by comparison with geometric series. A special test involving integrals was developed as a tool for investigating how fast a series converges, i.e. how many terms are needed to provide an estimate for the infinite sum with a specified measure of accuracy.

We introduced absolute convergence in the context of complex numbers, although it is a useful concept for real series. We presented a remarkable example of a situation where changing the order of summation changes the sum. Finally, we considered power series, which are used as tools for working with approximations to functions.

The chapter has involved some analytical proofs, in particular that of the existence of the radius of convergence of a power series, using the least upper bound axiom for the real number system. Some analytical reasoning also occurred in establishing the validity of the tests for convergence. The most important aspect to the chapter however is the examples, which have attempted to show how to apply the various results, and the exercises giving practice in such applications.

EXERCISES ON CHAPTER 8

1. Prove the formulae obtained in Example 2 by mathematical induction, using trigonometric identities but without reference to complex numbers.

2. Determine the convergence or otherwise of the series Σa_n, where a_n is defined by the following formulae.

   (i) $\dfrac{n+1}{n^2-3}$, (ii) $\dfrac{n-2}{(-3)^n}$, (iii) $\dfrac{2^n(n^3+1)}{5^n(n+4)}$, (iv) $\dfrac{(-1)^n \ln n}{\sqrt[7]{n}}$.

3. Estimate the number of terms needed to evaluate $\Sigma n^{-\pi}$ correct to within 10^{-5}.

4. Find the radius of convergence for each of the following power series. Show that the first converges everywhere on the circle of convergence, and that the second diverges everywhere on the circle of convergence.

   (i) $\displaystyle\sum_{n=0}^{\infty} \dfrac{(-1)^n z^n}{(n+1)^2}$, (ii) $\displaystyle\sum_{n=0}^{\infty}(-1)^n(2^n+n^2)z^n$.

5. Adapt the method used in Example 14(iv) to prove that if $0 < k \le a_n \le K$ for all $n \in \mathbb{N}$ then the power series $\Sigma a_n z^n$ has radius of convergence equal to 1, and that it diverges everywhere on the circle of convergence.

9 • Decimals

Decimals are a familiar topic in school mathematics. They have become more important in recent years with the introduction of the decimal system of money in the UK in 1971, and an increasing use of the metric system for measurement. They permit an ease of calculation which the previous systems of measurement used in the UK and the USA do not have (although those have advantages such as divisibility of many of the units by 3).

We shall not be concerned with the arithmetic of decimals in this chapter. We shall instead concentrate on their analysis, especially in relation to infinite decimals. As a particular case we shall analyse some of the properties of recurring (periodic) decimals.

The positional significance of the digits is what gives the decimal notation its essential structure, and this is learned at a very early stage in relation to integers, in interpreting the meaning of the separate digits in a number like 362, whereas the structural meaning of the decimal notation is learned rather later. For example, we can interpret 3.125 as

$$3 + \frac{1}{10} + \frac{2}{100} + \frac{5}{1000} = 3\frac{125}{1000} = 3\frac{1}{8} = \frac{25}{8},$$

where the last three steps are using the arithmetic of fractions. We learn something of this process at school, and also the reverse process of converting a fraction into a decimal equivalent. When we convert a fraction to a decimal we encounter examples like 1/3, where the conversion algorithm gives a potentially infinite sequence of threes. We shall consider what this means, along with similar phenomena.

In using decimals for measurement we often use the convention that a decimal does not represent an exact number, since there will always be errors of measurement. So, if for example we said that a length of wire measured 2.17 metres, we would conventionally mean that its true length was somewhere between 2.165 and 2.175 metres. This then leads to an analysis of the way in which such measurement errors combine. We will not consider this aspect of decimals in this book.

In the decimal system we use powers of 10 as denominators. This is based on a long history, and is, of course, related to the number of fingers most humans have, hence the use of the word 'digits'. Mathematically, there is no reason why the number 10 should not be replaced by an arbitrary positive integer r. We could for example use 6 in place of 10. In that framework we would intepret 3.125 as

$3 + \frac{1}{6} + \frac{2}{6^2} + \frac{5}{6^3} = 3\frac{51}{216} = \frac{699}{216}$. In practice, the use of 6 has little application, but replacing 10 by 2, 8 or 16 is of use in connection with computer arithmetic, as was mentioned in §2.6. The terminology is somewhat confused here. Some authors will talk about 'decimals in base six', others will use terms like 'siximals' or 'heximals'. We shall talk about 'expansions in base six', since much of our discussion will concern expanding a number into digital form.

9.1 Infinite Decimal Expansions

In this section we shall consider some of the theory behind decimal expansions, and link it with the work on sequences and series in the previous two chapters. We shall present the topic in a way which would enable 10 to be replaced with a general positive integer r as discussed at the end of the introduction. We prefer to use 10 as the base in the exposition, rather than a general value of r, as that will help to keep the theory in touch with the decimals which readers will be accustomed to.

An infinite decimal is an expression of the form $a_0 + \dfrac{a_1}{10} + \dfrac{a_2}{10^2} + \dfrac{a_3}{10^3} + \cdots,$

i.e. an infinite series, where the numbers (a_n) satisfy $0 \leq a_n < 10$ for all $n \geq 1$. The expression above is an infinite series with non-negative terms, and we have $a_n/10^n \leq 9/10^n$. The right-hand side of this inequality is the nth term of a geometric series with common ratio $1/10 < 1$, which is therefore convergent. The original series is therefore convergent by comparison. This establishes that every infinite decimal represents a real number. We know already that finite decimals represent real numbers just from the arithmetic of fractions. The next task is therefore to investigate the converse, i.e. if we start with a real number is there a decimal expansion for that number, and can we find it? What we shall do is to develop an algorithm which starts with a real number and constructs an expansion. We then have to prove that the sum of the resulting infinite series is equal to the number we started with.

When we convert a fraction to a decimal we do it by division. We shall analyse this process and convert it to an algorithm based on integer arithmetic. We consider the example 297/43, for which the first few steps of the long division process appear as

$$
\begin{array}{r}
6.906\ldots \\
43\,\overline{)\,297.000\ldots} \\
\underline{258} \\
390 \\
\underline{387} \\
300 \\
\underline{258} \\
420\ldots.
\end{array}
$$

The first step is to express the fraction 297/43 in the form of a quotient and a remainder, i.e. $297 = 6 \times 43 + 39$. This is reminiscent of an individual step in the Euclidean Algorithm (§2.5). The next step is to multiply the remainder by 10 and again divide by 43. (Multiplying by 10 corresponds to 'bringing down a zero' in long division.) This process continues, and so we can write out the steps exhibited in the long division as follows.

$$297 = 6 \times 43 + 39$$
$$10 \times 39 = 9 \times 43 + 3$$
$$10 \times 3 = 0 \times 43 + 30$$
$$10 \times 30 = 6 \times 43 + 42.$$

This is now in a form which can be translated into a simple computer program, a possible core of which is the following segment of Pascal, where we start with a and b playing the roles of 297 and 43 respectively. The program is typical of others we have met in this book, involving a repeat -- until loop, which reflects what we have described mathematically as an iterative procedure. The similarity of structure with the programs sketched in Examples 7 and 12 of Chapter 2 should be noted.

```
repeat
        quotient:= a div b;
        remainder:= a mod b;
        write(quotient);
        a:= 10*remainder;
until rem=0 or (some stopping condition).
```

The program obviously needs adjusting in respect of line 4 so that the decimal point appears, and some thought needs to be given to a stopping condition, for as we shall see such a program could otherwise continue for ever. The easiest method would probably be to insert a variable which increases by 1 each time the loop is repeated, and stops the program when it reaches a specified value, i.e. after a chosen number of steps. Notice that in the steps of the method and the program we could replace 10 by 6 or some other number r to obtain the expansion in another base. To understand this we note that another way of looking at a step such as $10 \times 39 = 9 \times 43 + 3$ is that it determines how many tenths there are in $39/43$. Replacing 10 by 6 would therefore determine how many sixths there are in $39/43$.

TUTORIAL PROBLEM I

Expand the Pascal segment above into a working program, and test it. (A similar structure in Basic will work.)

We now use the idea from the programming approach to formulate the algorithm in mathematical terms, using notation that will identify the quotients with the digits in the expansion. We shall want the algorithm to be able to cope with non-rational numbers and so we cannot use division as we have with fractions. Another way to think about the quotient when a is divided by b is to regard it as the largest integer which does not exceed $x = a/b$. This is a well-known function of a real variable, denoted by $[x]$. It is also a function of a real variable in Pascal, denoted by `int(x)`. We shall use the latter notation in the mathematical development, as it gives a clearer reminder of the integer part function than the more abbreviated $[x]$.

• Proposition I

To convert a real number into a decimal expansion.   ●

Algorithm

Given a real number x, let $a_0 = \text{int}(x)$ and let $x_1 = x - a_0$, so that $x = a_0 + x_1$. We now define sequences (a_n) and (x_n) inductively by $a_{n-1} = \text{int}(10 \times x_{n-1})$ and $x_n = 10 \times x_{n-1} - a_{n-1}$. We then have $0 \le x_n < 1$ for all n, since x_n is the result of removing the integer part from a real number. So $0 \le 10 \times x_n < 10$ and therefore $0 \le a_n < 10$ for $n \ge 1$ because $a_n = \text{int}(10 \times x_n)$.

Finally we have $x = \displaystyle\sum_{n=0}^{\infty} \frac{a_n}{10^n} = a_0.a_1 a_2 a_3 \ldots$ in decimal notation.   ●

VERIFICATION The first thing to note is that because $0 \le a_n < 10$ for $n \ge 1$, the infinite series converges, as explained at the beginning of this section. To show that the sum of the series is equal to x we need to consider the sequence of partial sums. We prove that

$$0 \le x - \sum_{n=0}^{m} \frac{a_n}{10^n} < \frac{1}{10^m}.$$

For $n = 0$ this statement becomes $0 \le x - a_0 < 1$, which is true. In order to understand the algorithm better we shall work through the next two steps, even though, for an inductive proof, only the initial step is logically required.

So we have $a_1 = \text{int}(10x_1)$, and we can write $10x_1 = a_1 + x_2$ where $0 \le x_2 < 1$. We deduce that

$$x_1 = \frac{a_1}{10} + \frac{x_2}{10} \quad \text{giving} \quad x = a_0 + \frac{a_1}{10} + \frac{x_2}{10} \quad \text{where} \quad 0 \le x_2 < 1.$$

We now execute the second step, so $a_2 = \text{int}(10x_2)$, and we can write $10x_2 = a_2 + x_3$ where $0 \le x_3 < 1$. We deduce that

$$x_2 = \frac{a_2}{10} + \frac{x_3}{10} \quad \text{giving} \quad x = a_0 + \frac{a_1}{10} + \frac{a_2}{10^2} + \frac{x_3}{10^2} \quad \text{where} \quad 0 \le x_3 < 1.$$

We now perform the inductive step, so suppose that

$$x = a_0 + \frac{a_1}{10} + \frac{a_2}{10^2} + \frac{a_3}{10^3} + \ldots + \frac{a_{m-1}}{10^{m-1}} + \frac{x_m}{10^{m-1}} \quad \text{where} \quad 0 \le x_m < 1.$$

The algorithm then gives $a_m = \text{int}(10x_m)$ and so $10x_m = a_m + x_{m+1}$, where $0 \le x_{m+1} < 1$. We can then substitute for x_m to obtain

$$x = a_0 + \frac{a_1}{10} + \frac{a_2}{10^2} + \frac{a_3}{10^3} + \ldots + \frac{a_{m-1}}{10^{m-1}} + \frac{a_m}{10^m} + \frac{x_{m+1}}{10^m} \quad \text{where} \quad 0 \le x_{m+1} < 1.$$

From this equation we conclude that for all $m \in \mathbb{N}$,

$$0 \le x - \sum_{n=0}^{m} \frac{a_n}{10^n} = \frac{x_{m+1}}{10^m} < \frac{1}{10^m}.$$

This tells us that as m increases the difference between x and the partial sum of the series tends to zero, so that the infinite series converges, with sum x.   ●

Two questions arise immediately as a result of this algorithm. Firstly, can any sequence of digits (a_n) arise from the algorithm? Secondly, can a number x have more than one decimal representation, possibly obtained through a different algorithm? The answer to the second question is yes, but in one sense only. If we consider the decimal $0.999\ldots$ having an infinite sequence of nines, this represents a geometric series whose first term is $9/10$ and whose common ratio is $1/10$. The series therefore converges, and the formula for the sum of an infinite geometric series gives the answer 1. This implies that the decimals $1.000\ldots$ and $0.999\ldots$ represent the same number. A similar example which readers can verify is that the decimals $0.25000\ldots$ and $0.24999\ldots$ both represent the number $1/4$. We shall discuss this further in the last section of this chapter, but at this point we shall prove that this is the only multiple representation which can occur, so that apart from this kind of case we will have uniqueness.

We first demonstrate that the algorithm itself will never give rise to an infinite sequence of nines. Suppose a number x corresponds to a decimal for which $a_n = 9$ for all $n > K$. Then

$$x - \sum_{n=0}^{K} \frac{a_n}{10^n} = \sum_{n=K+1}^{\infty} \frac{9}{10^n} = \frac{1}{10^K},$$

summing the infinite geometric series. But we have shown that for decimals arising from the algorithm we have $x - \sum_{n=0}^{K} \frac{a_n}{10^n} < \frac{1}{10^K}$, so that equality is impossible.

We now show that apart from an infinite sequence of nines we cannot have two different decimal representations for a given number. This entails an affirmative answer to the first question above, for if we have a decimal we know that it converges to some number x, and this uniqueness result tells us that we could not obtain a different sequence of digits from the algorithm applied to x. The proof is by contradiction, so let us suppose that

$$\sum_{n=0}^{\infty} \frac{a_n}{10^n} = \sum_{n=0}^{\infty} \frac{b_n}{10^n}.$$

Suppose that the sequences (a_n) and (b_n) differ for the first time when $n = K$, and that without loss of generality $a_K > b_K$. Then

$$0 = \sum_{n=0}^{\infty} \frac{a_n}{10^n} - \sum_{n=0}^{\infty} \frac{b_n}{10^n} = \frac{a_K - b_K}{10^K} - \sum_{n=K+1}^{\infty} \frac{b_n - a_n}{10^n} \ge \frac{1}{10^K} - \sum_{n=K+1}^{\infty} \frac{b_n - a_n}{10^n}.$$

Now $-9 \le b_n - a_n \le 9$ for all n, and unless $b_n = 9$ and $a_n = 0$ for all $n \ge K+1$ the right-hand sum will be strictly less that the geometric series

$$\sum_{n=K+1}^{\infty} \frac{9}{10^n}, \quad \text{whose sum is} \quad \frac{1}{10^K}, \quad \text{implying} \quad 0 \ge \frac{1}{10^K} - \sum_{n=K+1}^{\infty} \frac{|b_n - a_n|}{10^n} > 0,$$

which is a contradiction. This shows that if we exclude infinite sequences of nines, the decimal representation is unique, and when we speak of *the* decimal representation we shall be referring to the one not containing an infinite tail of nines.

Infinite tails of other digits are not ruled out, and one particular case is where the digits are all zero from some point onwards. Essentially this means that we can consider the algorithm as finishing in a finite number of steps, since all the x_n are zero from some point onwards. We refer to such decimals as terminating or finite decimals. These correspond to fractions of a particular form as we shall now demonstrate. If we first take a finite decimal $a_0.a_1a_2 \ldots a_m$, this represents

$$a_0 + \frac{a_1}{10} + \frac{a_2}{10^2} + \ldots + \frac{a_m}{10^m} = \frac{a_0 10^m + a_1 10^{m-1} + a_2 10^{m-2} + \ldots + a_m}{10^m}.$$

This is a fraction whose denominator is a power of 10. There may be some cancellation possible, but after this has occurred the resulting denominator will still not have any prime factors apart from 2 and 5. Conversely, suppose we have a fraction whose denominator is of the form $2^p 5^q$. We can convert it into a fraction whose denominator is a power of 10 by multiplying numerator and denominator by a power of either 2 or 5, so as to make the two powers the same. A fraction with a power of 10 as its denominator then clearly gives a finite decimal, since the digits of the numerator will be those of the decimal itself. We can illustrate this with an example,

$$\frac{7493}{16000} = \frac{7493}{2^7 5^3} = \frac{7493 \times 5^4}{2^7 5^3 \times 5^4} = \frac{4683125}{10^7} = 0.4683125.$$

This tells us that if we have a fraction in its lowest terms whose denominator contains a prime factor different from 2 or 5 then its decimal representation will not terminate.

The proof of Proposition 1 and the subsequent discussion about uniqueness work in exactly the same way if we consider expansions in base r in place of base 10. Infinite sequences of nines would become infinite sequences of $(r-1)$s; so, for example, in base 6 the expansion $0.31555\ldots$ and $0.32000\ldots$ represent the same number.

TUTORIAL PROBLEM 2

Show that in base r a fraction in its lowest terms has a terminating expansion if and only if the denominator has no prime factors which are not prime factors of r.

9.2 Periodic Decimals

If we convert some fractions into decimals using the division algorithm explained in the previous section we find some interesting patterns of digits appearing. In using examples to illustrate these patterns the arithmetic calculations will be left to the reader. Most pupils at school encounter the decimal for $1/3$, where the division process gives a potentially unending sequence of threes. A few other examples demonstrate the possibilities:

$$\frac{3}{11} = 0.2727\ldots, \quad \frac{5}{24} = 0.208333\ldots, \quad \frac{23}{75} = 0.30666\ldots,$$

$$\frac{5}{7} = 0.714285714285714\ldots$$

In each of these examples we find a repeating pattern. In some it starts at the beginning and in others after a few initial digits. Decimals with this repeating property are called periodic, or recurring. It is clearly inconvenient to write down the periodic sequence several times and so there are various notations to indicate this, the most common of which consists of a dot placed over the recurring digit when there is just one, or the first and last digits of the period otherwise. In this notation we would write

$$\frac{3}{11} = 0.\dot{2}\dot{7}, \quad \frac{5}{24} = 0.208\dot{3}, \quad \frac{23}{75} = 0.30\dot{6}, \quad \frac{5}{7} = 0.\dot{7}1428\dot{5}.$$

● *Proposition 2*

Every fraction gives rise to a decimal which is eventually periodic. If the denominator is n then the length of the sequence of repeating digits is at most $n - 1$.

●

PROOF

We suppose that the integer part of the fraction has been determined, and we will therefore restrict the discussion to fractions between 0 and 1, so that the numerator is less than the denominator. If we recall the division process applied to a fraction m/n we can see that at each stage the remainder is multiplied by 10 to become the numerator for the subsequent division. If we obtain a remainder of zero then the process has terminated and we obtain a finite decimal. In terms of periodicity we could consider this as having an infinite sequence of zeros, and therefore having period length 1. If we do not obtain a remainder zero the division process continues without ending. In that case the number of different remainders is at most $n - 1$, and as soon as we obtain a remainder which has occurred previously the sequences of quotients and remainders begin repeating. We therefore have periodicity, of length at most $n - 1$.

●

● *Example 1*

Find the decimal expansion for $1/17$, using a calculator.

We shall modify the division algorithm and use the calculator to generate several digits at once. The calculations which follow can be set out on paper to mimic long division. The numerical results of course refer to the author's own ancient calculator. Readers should follow this example with a calculator. Entering $1 \div 17$ gives 0.058823529. We shall use all but the last digit, keeping that as a check. The multiplication 5882352×17 gives 99999984. Subtracting this from an appropriate power of 10 will give a remainder of 16. So the next step involves $16 \div 17$. This gives 0.94117647, and we can see that the first digit 9 agrees with the 9 which we retained as a check from the first step. We now multiply 94117647×17 and obtain 1599999999, which leaves remainder 1 when subtracted from $16 \times$ a power of 10. We began the process by dividing 1 by 17, and so we will have repetition from this point on. In fact we have generated 16 digits in these two steps, and we know that the period length cannot be more than 16. We can summarize these calculations in a form which reflects the division algorithm as follows:

$$100000000 = 5882352 \times 17 + 16; \quad 1600000000 = 94117647 \times 17 + 1.$$

Finally we can write $1/17 = 0.\dot{0}58823529411764\dot{7}$, which has period length 16.

In other cases the period will be less than $n - 1$, as in the example 3/11, which has period 2. To calculate the length of the period without working out the decimal itself we can proceed as follows. We first express the fraction in its lowest terms. We then remove from the denominator any powers of 2 or 5, giving a number M. In general, the length of the period depends on the remainders when successive powers of 10 are divided by M. The smallest power which has remainder 1 gives the length of the period. We can illustrate this with some examples. For those quoted above we find that for 5/24 we have $24 = 2^3 \times 3$, so that $M = 3$, and 10^1 has remainder 1 on division by 3, giving period length equal to 1. For 5/7 successive powers of 10 give remainders as follows when divided by 7

$$10 : \text{rem } 3, \ 10^2 : \text{rem } 2, \ 10^3 : \text{rem } 6, \ 10^4 : \text{rem } 4, \ 10^5 : \text{rem } 5, \ 10^6 : \text{rem } 1.$$

So 5/7 has period length 6. For 3/11, 10 gives remainder 10 on division by 11 (with quotient zero) and 100 gives remainder 1, so that 3/11 has period length 2. The general result can be expressed as follows, but the proof is outside the scope of this book.

● *Proposition 3*

Consider the decimal corresponding to the fraction a/b, where a and b have no common factors. If $b = 2^p 5^q$ and $\max(p, q) = m$ then the decimal terminates after m digits. If $b = 2^p 5^q M$, where $M > 1$ and M is not divisible by 2 or 5, and if c is the smallest power of 10 giving remainder 1 on division by M then the decimal begins with m non-recurring digits and is then periodic with period length c. ●

TUTORIAL PROBLEM 3

Work out some examples to illustrate the result of Proposition 3.

If we have a decimal with denominator n and with the maximum period length $n - 1$ then the sequence of remainders must contain all the integers between 1 and $n - 1$ inclusive, for otherwise there would be repetition after fewer than $n - 1$ digits. So calculating the decimal for m/n simply corresponds to starting the division process at the position where m appears in the sequence of remainders. So all fractions of this form will have the same periodic sequence for the recurring part of the decimal, but starting at a different place. The most familar example concerns the fractions with denominator 7. If we write down the decimals for these fractions with the numerators in the order in which the remainders for powers of 10 appeared above we find that

$$\frac{3}{7} = 0.\dot{4}2857\dot{1}, \ \frac{2}{7} = 0.\dot{2}8571\dot{4}, \ \frac{6}{7} = 0.\dot{8}5714\dot{2}, \ \frac{4}{7} = 0.\dot{5}7142\dot{8},$$

$$\frac{5}{7} = 0.\dot{7}1428\dot{5}, \ \frac{1}{7} = 0.\dot{1}4285\dot{7}.$$

We have shown that a fraction corresponds to a recurring (or finite) decimal. We now prove the converse result.

● Proposition 4

Every recurring decimal corresponds to a fraction.   ●

PROOF

If we have a recurring decimal with a few initial digits we can decompose it as the following example illustrates

$$0.253\dot{4}7\dot{1} = 0.253 + 0.000\dot{4}7\dot{1} = 0.253 + 0.\dot{4}7\dot{1} \times 10^{-3}.$$

If we know that the decimal $0.\dot{4}7\dot{1}$ corresponds to a fraction, then since a finite decimal also corresponds to a fraction, the decomposition shows that the original decimal gives a fraction. This demonstrates that we need only prove the result for purely periodic decimals, i.e. those with no initial digits. So suppose we have the decimal $0.\dot{a}_1 a_2 \ldots \dot{a}_n$. If we consider the finite decimal consisting of just one period we have

$$0.a_1 a_2 \ldots a_n = \frac{a_1 10^n + a_2 10^{n-1} + \ldots + a_n}{10^n} = \frac{A}{10^n},$$

where A stands for the numerator. The recurring decimal therefore corresponds to a geometric series with common ratio $1/10^n$, which we can sum as follows

$$\frac{A}{10^n} + \frac{A}{10^{2n}} + \frac{A}{10^{3n}} + \ldots = \frac{A}{10^n - 1},$$

which is a fraction.   ●

Proposition 4 will also work if 10 is replaced by a general number r as the base of expansion, because the calculations in the proof are algebraic and not arithmetic in nature.

Example 2

Find the fractions corresponding to the decimals $0.4\dot{7}\dot{1}$ and $0.253\dot{4}7\dot{1}$.

We could use the general result in the proof of Proposition 4, but it gives a better understanding to work from first principles. The recurring decimal corresponds to a geometric series which we can sum, as follows

$$0.471471471\ldots = \frac{471}{10^3} + \frac{471}{10^6} + \frac{471}{10^9} + \cdots = \frac{471}{10^3}\left(1 + \frac{1}{10^3} + \frac{1}{10^6} + \cdots\right)$$

$$= \frac{471}{10^3}\frac{1}{1-(1/10^3)} = \frac{471}{10^3}\frac{10^3}{10^3-1} = \frac{471}{999} = \frac{157}{333}.$$

We can now follow the first part of the procedure in the above proof and write

$$0.253\dot{4}7\dot{1} = 0.253 + \frac{0.4\dot{7}\dot{1}}{10^3} = \frac{253}{1000} + \frac{157}{333 \times 1000} = \frac{84406}{333000}.$$

We can see from the first part of this example that there is an easily remembered rule for converting a purely periodic decimal into a fraction. The numerator consists of the integer whose digits form a single period, with length k say, and the denominator consists of k nines. So with this rule we can write $0.\dot{4}7\dot{1} = 471/999$, $0.\dot{2}\dot{7} = 27/99 = 3/11$, $0.\dot{0}023\dot{1} = 00231/99999 = 231/99999 = 77/33333$. Notice in the last example that we need to be careful when there are zeros at the beginning of the period. In case of doubt we can always use the geometric series.

TUTORIAL PROBLEM 4

Devise a rule for converting a purely recurring expansion in base 6 into a fraction. Illustrate your result with an example and verify it by using the geometric series.

A consequence of the propositions of this chapter is that any irrational number will have a decimal which is neither terminating nor recurring, so that the sequence of digits has no infinitely repeating pattern. In particular, numbers like $\sqrt{2}$, π, and the exponential number e have decimals which are infinite and non-periodic. The generalizations we have commented on relating to other bases of expansion tell us that irrational numbers do not have periodic expansions in any base. There are many interesting questions which can be asked about the overall proportion of individual digits or sequences of digits in such expansions, but they are outside the scope of this book.

In the case of fractions some bases will give finite decimals whereas others will give infinite periodic expansions. For example, in base 10 (decimal) we have $1/3 = 0.\dot{3}$, whereas in base 6 we have $1/3 = 2/6 = 0.2$, reflecting the fact that the expansion terminates if the base and the denominator do not have any prime factors different from one another.

> Investigate expansions in various bases for some fractions, illustrating in
> particular the comment above arising from the expansions of $1/3$.

9.3 Point Nine Recurring

The assertion that $0.\dot{9} = 1$ causes problems for many students. Giving a
mathematical explanation, as we did when discussing uniqueness in the previous
section, seems not to remove these doubts, and so we have to look for reasons which
are partly psychological. Most situations involving numbers exhibit some kind of
uniqueness, beginning with the standard digital representation of whole numbers in
base ten. It is inconceivable that a number such as 2314 could have an alternative
representation in base ten such as 3702. It is this desire for uniqueness which seems
to be part of the psychological difficulty. Learning non-uniqueness in the case of
rational numbers causes some pupils problems, and to recognize why 4/6 and 14/21
represent the same number is a non-trivial task, at whatever level one approaches
this. We demonstrated an approach using equivalence relations in Chapter 3, and
there are naturally explanations within the context of school mathematics. Many
introductions to the study of decimals are closely related to the division process,
where uniqueness is guaranteed. This comes out of experience, so if you repeat the
process with a given fraction, a different decimal will not result except from an
arithmetical error. In fact, we showed above that the ambiguous case of an infinite
tail of nines could not occur through the algorithm. In areas other than mathematics
there is a constant search for uniqueness of representation so as to avoid ambiguity
of communication. In the sciences, highly technical language is developed with this
aim in mind. An example from geography concerns latitude and longitude. It is
sometimes difficult to remember that latitude goes from 180°E to 180°W whereas
longitude goes from 90°S to 90°N. Why couldn't they both use the same interval? In
fact, if both used 180° for the two extremities each position on the surface of the
earth would have two sets of coordinates, for example the point 10°W, 50°N would
be the same as 170°E, 130°N. A choice of interval also occurred when we specified
the argument of a complex number in Chapter 6. Similar choices occur when we
describe cylindrical and spherical polar coordinates in three-dimensional geometry.
Sometimes we learn to tolerate multiple representation. For example, in dealing with
the complex exponential, as in §6.5, it is so useful to allow θ to be unrestricted when
we are using $e^{i\theta}$ that it is worth paying the price of non-uniqueness.

Returning to the question of $0.\dot{9}$, acceptance of this as a legitimate decimal saves
having to mention a condition such as 'provided the decimal contains infinitely
many digits different from 9' every time we undertake some general exploration of
decimals. We decide to allow all sequences of digits, and then recognize that
numbers having terminating decimals possess the alternative expansion with a tail of
nines. This is a mathematical decision, and will not necessarily be expected to

remove the psychological feeling that $0.\dot{9}$ ought somehow to be just slightly less than 1. We have to recognize that the decision to accept equality is made for mathematical reasons, and come to terms with the intellectual consequences. This is a topic which has engaged the interest of many teachers in schools and universities over the years, and you may be interested in some articles on the subject which have appeared from time to time in the journal *Mathematics Teaching*. Issue No. 115 (June 1986) contains one such article, and it refers to others. A collection appeared in *Readings in Mathematical Education—Exploring Numbers* published by the Association of Teachers of Mathematics in 1987.

Summary

This chapter contains no end-of-section exercises. The aim has not been to develop purely arithmetical skills in dealing with decimals, indeed it is hoped that the calculations involved in examples will be done by readers using a calculator or a computer. A study of the material of the chapter will indicate that there is more to decimals than simple arithmetic tools. Their structure is based on that of infinite series, and therefore depends ultimately on the analysis of limiting processes, unlike the study of fractions, which is less analytic and more algebraic in nature. The results proved in the chapter are designed to gather together the basic facts about decimal representations, in particular the notion that rational numbers have decimals which are either finite or recurring, and irrational numbers have decimals which are infinite and not periodic. The other theoretical aspect studied is an analysis of the standard algorithm for constructing decimals, which in the case of fractions is considered from the perspective of division. Finally, we have looked briefly at patterns which can occur in the periodic representation of fractions. This is a fascinating topic, and we hope that the small number of illustrations will have whetted the reader's appetite to carry out further investigations.

EXERCISES ON CHAPTER 9

1. Use your calculator as in Example 1 to find the decimal expansion for $1/47$.

2. Investigate the cyclic patterns occurring within the decimals for fractions with denominator 13.

3. Without working out the decimal expansions, determine the period lengths of the decimal expansions for the fractions $3/21$, $12/43$, $53/96$, $7/23$.

4. Write a computer program to find the period length of the decimal expansion of a fraction m/n.

10 • Further Developments

In this final chapter we discuss briefly a few of the ways in which the topics in the previous chapters progress further, some of which will be met in later undergraduate courses. We shall give references to further reading and make some suggestions for self-study projects which students can undertake, with the help of a tutor, if that option is available within their course. In a few places we have referred to the historical context for mathematical ideas, but we have barely scratched the surface of that aspect, and so we urge readers to delve into the history of the subject from time to time, and to that end we recommend C.B. Boyer and U.C. Merzbach, *A History of Mathematics* (2nd Edition), John Wiley & Sons, 1989. For books concerned more specifically with the history of numbers we recommend *Numbers, their History and Meaning* by Graham Flegg, Pelican, 1984 and *Numbers and Infinity* by E. Sondheimer and A. Rogerson, Cambridge University Press, 1981.

10.1 Sets, Logic and Boolean Algebra

The discussion of logic in Chapter 1 was undertaken chiefly to provide some background for the study of the reasoning involved in mathematical proofs, rather than as a topic in its own right. We did not therefore study logical symbolism and its uses in detail. Symbolic logic is an interesting topic and enables one to analyse the structure of complicated statements with many components by means of algebraic symbolism and calculations. Logical puzzles can also be solved using these tools. Statements are either true or false, i.e. they have two possible states. The elementary components of the circuits used in computers also have two states, on and off, just like switches. The study of such systems is associated with the name of George Boole (1815–1864), whose best known work *Investigation of the Laws of Thought* was published in 1854. Another direction for the study of sets and logic was the questioning of foundations, developing systems of axioms for sets themselves. This is a somewhat abstract and esoteric topic, but it occupied many eminent mathematicians in the years either side of 1900. One of the best known results, which arose out of this perspective, is Gödel's Incompleteness Theorem, which essentially says that any logical system extensive enough to contain arithmetic will have statements which can neither be proved nor disproved within the system. There are many popular accounts of this spectacular result, and we refer the reader to the splendid book by Douglas R. Hofstadter, *Gödel, Escher, Bach*, Harvester Press 1979, for this and many other insights into mathematics. For the reader who is interested in some of the philosophical issues we recommend *Language, Logic and Mathematics* by C.W. Kilmister, English Universities Press, 1967.

Project 1

A study of Boolean Algebra with applications to switching circuits. There are many books which cover this topic, but we recommend Chapter 1 from the second or subsequent editions of the classic book, first published in 1956, by J.G. Kemeny, J.L. Snell and G.L. Thompson, *Introduction to Finite Mathematics*, Prentice Hall.

One of the far-reaching discoveries of the late 19th century was made by Georg Cantor (1845–1918), who invented methods of comparing the sizes of infinite sets. This essentially enables us to say that two sets have the same size by matching them via a one-to-one correspondence. For example, I can tell that I have the same number of fingers on both my hands without counting them, simply by physically matching them together. When we do this for infinite sets we begin to encounter properties which are radically different from ordinary finite sets. For example, the function $f(n) = 2n$ describes a matching between the set of all integers and the set of even integers, so that in Cantor's theory an infinite set and some of its subsets might be equally numerous. Because this is so different from our experience with finite sets it is sometimes referred to as a paradox, but it is logically valid. Cantor was able to develop theories of infinite numbers on this basis, and to investigate their arithmetic.

Project 2

A study of Cantor's theories of cardinal and ordinal number. As a starting point we suggest an article by the author of this book *Infinite sets and how to count them* which appeared in the journal *Mathematics Teaching*, No. 58, 1972. A highly entertaining book which sets the theory in the context of improbable stories such as a hotel with an infinite number of rooms is *Stories about Sets* by N.Y. Vilenkin, Academic Press, 1968. The book *Numbers and Infinity* by Sondheimer and Rogerson cited in the introduction also touches on this topic.

10.2 Number Theory and Continued Fractions

To Gauss is attributed the remark 'Mathematics is the queen of the sciences, and number theory is the queen of mathematics'. In terms of historical antiquity, number theory is on a par with geometry, for the books of Euclid concern both topics. In Chapter 2 we met the Euclidean Algorithm, a remarkable result which has proved to be an efficient calculating tool for over 2000 years. Number theory has been at the root of many branches of mathematics over the centuries, and many university courses will include a study of parts of the topic. For students who are sufficiently fascinated by numbers to wish to begin a study of the subject on their own we recommend *A Pathway Into Number Theory* by R.P. Burn, Cambridge University Press, 1982. This book is organized as a progression of problems through which readers can build up an insight into number theory at their own pace.

Project 3

A study of solving equations in modular arithmetic. We have met the idea of remainders in several places in this book, especially when dealing with decimals. Modular arithmetic is the arithmetic of remainders, so that if we take, for example, the basic divisor (modulus) as 7, then as far as remainders are concerned, we can say that $5 + 4 = 2$. This gives rise to a finite arithmetic, for the only possible remainders are $0 \ldots 6$. We can investigate solving both linear and quadratic equations in such finite systems.

In Chapter 3 we introduced continued fractions. This is a topic which is important in number theory, but is not always included in an introductory course. We studied continued fractions for rational numbers, but irrational numbers also can be expanded in that form, and they are used for approximations. A good introduction is the book *Continued Fractions* by C.D. Olds. This is part of the New Mathematical Library, published by the Mathematical Association of America. The series as a whole is well worth exploring.

Project 4

A study of continued fractions for quadratic irrationals, i.e. numbers of the form $(p + \sqrt{d})/q$ where p, q, d are integers. The theory, established by Euler, Lagrange and others, shows that such quadratic numbers have continued fractions which are periodic. Of particular interest are those associated with square roots such as $\sqrt{76}$, and there are many patterns within these numbers which students can investigate for themselves.

10.3 Real Numbers and More

In Chapter 5 we gave a brief historical account of the background leading to the development of theories of real numbers at the end of the 19th century. That historical process rejected the notion of infinitely small quantities with which Newton had sought to reason. The matter seemed to have been settled by Dedekind and Cantor, but in 1960 the American mathematician Abraham Robinson resurrected the idea, this time by establishing a logically rigorous formulation of a hyperreal number system, having the ordinary real numbers embedded as a subsystem, and having well-defined arithmetic operations and an order relation, but containing additional elements, called infinitesimals, which were different from zero but less that any ordinary positive real number. This system then was able to serve as a logical foundation for the theory of limits in which, broadly speaking, the component 'for all $\epsilon > 0$' which we have used in the definition of limit is replaced by 'for all infinitesimals'. This method of approach to the calculus has not caught on, and present day undergraduate courses in analysis take essentially the point of view developed by Cauchy and Weierstrass in the 19th century, with the theory of real numbers included. Interestingly, the new development is referred to as 'non-standard' analysis, and it features in final year courses only in some universities.

Project 5

One of the chief advocates of a 'non-standard' approach to calculus is H.J. Keisler, who published a book called *Elementary Calculus*, Prindle, 1976. This can form the basis of an interesting final year project, with tutorial supervision.

10.4 Complex Numbers and Beyond

We saw in Chapter 2 that Hamilton discovered in 1833 how to represent the complex numbers as an algebra of pairs. He spent a good deal of time subsequently searching for a way of formulating a system of triples (a, b, c) which would obey the normal rules of algebra. He failed in this, largely because it is impossible! There is a short article which explains this by Kenneth O. May, *The Impossibility of a Division Algebra of Vectors in Three Dimensional Space,* which appeared in the *American Mathematical Monthly*, Vol. 73, 1966, pp 289–91. It was 10 years later in 1843 that Hamilton suddenly realized that a four-dimensional algebra was possible, and it is asserted that in his excitement he scratched the defining equations on a stone bridge. The system he invented is called the quaternions, and a brief introduction is given in Chapter 26 of the history of mathematics book recommended at the beginning of this chapter. Nowadays quaternions are most commonly met in courses on group theory or abstract algebra.

Returning to complex numbers, one of the topics we did not discuss in Chapter 6 was the geometry of complex functions. For a real function given by $y = f(x)$ both variables can be represented on a number line. They are therefore one-dimensional, and so the functional relationship between x and y can be represented in two dimensions through the Cartesian graph. For complex variables, with an equation like $w = z^3$, both variables are two dimensional, and so in theory we would need a four-dimensional space to draw a graph. To overcome this difficulty we proceed by considering two representations of the complex plane side by side. One of them corresponds to z and the other to w. If we therefore plot a point such as $z = 1 + i$, this will give rise to the point $(1 + i)^3 = -2 + 2i$ in the w-plane. If we imagine all points on the imaginary axis in the z-plane this corresponds to $z = yi$ with y varying over all real numbers. We then have $w = (yi)^3 = -y^3 i$. So w also lies on the imaginary axis. The negative sign indicates that if a point moves up the imaginary axis in the z-plane then its image in the w-plane will move *down* the imaginary axis there. Also, we see that the rate at which these lines are traced out will vary, for if y varies at a uniform rate then the rate of change of y^3 will be variable. One can investigate the images, in the w-plane, of many types of curves and regions in the z-plane under transformations defined by functions of z. Of particular importance are the bilinear or Möbius transformations defined by $w = (az + b)/(cz + d)$ where a, b, c, d are complex numbers. Complex transformations have many applications, for example in aerodynamics and in heat conduction problems. Investigating the algebra of transformations provides tools to construct the visual images, but now we have the added power of computer graphics. There are several packages available, among which is one of the options within Graphical Calculus, which we have referred to on other occasions in this book.

Project 6

A study of the geometrical aspects of complex transformations. Many books on complex analysis contain relevant material with varying emphasis. For one with a good coverage we suggest R.V. Churchill and J.W. Brown, *Complex Variables and Applications*, McGraw Hill. The first version of this was published in 1948, but it has been through several editions. Chapter 7 of the fourth edition (1984) for example discusses many of the standard transformations and is well illustrated.

10.5 Sequences and Series

Infinite series are used in many places in mathematics, and will be encountered in many university course units. They are used to represent functions in calculus, algebra, number theory, applied mathematics, theoretical physics, probability, statistics, etc. The accuracy of approximations derived by truncating series is discussed in Numerical Analysis. Approximations to solutions of differential equations which do not have exact algebraic solutions are often obtained by a set of methods known as 'solution in series', and sometimes this involves the sequence of coefficients in a power series being defined inductively through the differential equation. In probability theory and in combinatorial mathematics, series are used in the context of generating functions, where the aim is to use a very simple formula whose expansion in a power series will generate an infinite sequence of coefficients which may represent probabilities, or numbers of ways of counting collections of objects etc. Because they occur so widely we shall focus here on less common aspects of series.

The first concerns series which diverge. Euler himself thought that the series $1 + 1 - 1 + 1 - 1 + 1 - \ldots$, which is divergent, ought somehow to be associated with the number $\frac{1}{2}$, because that is the average of 0 and 1. This is a somewhat metaphysical way of arguing, but during the 19th and early 20th centuries several mathematicians developed theories which put the notion of associating a number with a divergent series through averages on a proper mathematical footing.

Project 7

The English mathematician G.H. Hardy wrote a book *Divergent Series*, first published in 1949. A study of this topic is rather specialized, but the first two chapters of that book give an excellent historical introduction to the subject, and contain some examples which can be studied with some profit. A further study of divergent series is probably better approached through somewhat older books, for example T.J. Bromwich, *Introduction to the Theory of Infinite Series*, which first appeared in 1908 but has been reissued many times. K. Knopp's *Theory and Applications of Infinite Series* is another book with a similarly distinguished history.

Both are rich in historical references to original sources. Books of this era contain other interesting aspects of series which are now studied very little.

Project 8

As well as adding an infinite sequence of numbers one can imagine multiplying a sequence, giving an infinite product. This can be approached through partial products, analogous to partial sums, and the limit of the sequence of partial products can be investigated. Books such as Bromwich and Knopp will generally have a section on infinite products, as does E.C. Titchmarsh, *The Theory of Functions*, Oxford University Press, 1932.

Project 9

Of rather more central importance is the topic of Fourier Series, which in some cases will be the subject of all or part of a lecture course. Even in those circumstances it is a sufficiently rich theme to provide subjects for individual projects. Most of the books mentioned above have a section on Fourier Series. Such series contain a real variable and can represent a function. Instead of powers like x^n the series are based on the sequences of trigonometric functions $(\cos nx)$ and $(\sin nx)$. They can therefore be used to provide good approximations to periodic functions, and the application to waveforms from physics and electronics is what makes them so useful, in addition to their mathematical importance. We referred to them briefly in §5.2. They have a central importance in the historical development of analysis during the 19th century, and a good account of this is to be found in J.H. Manheim *The Genesis of Point Set Topology*, Pergamon, 1964.

10.6 Decimals

Project 10

In Chapter 9 we gave a glimpse of some of the fascinating number patterns which appear in the study of periodic decimals. This can form the basis of a project which might be somewhat more exploratory and less book-based than some of those offered above. Writing programs to implement algorithms for converting fractions into decimals in various bases will help to reinforce understanding of the theory, and can then be used to generate data in order to look for patterns, make conjectures and attempt proofs.

Project 11

We hinted in Chapter 9 that even for non-periodic decimals there are interesting questions to be asked about the distribution of digits which can occur. Some aspects of this are discussed in G.H. Hardy and E.M. Wright, *The Theory of Numbers*, a classic in its field, published by Oxford University Press, first in 1938. In their chapter on decimals they discuss the intriguing result which states that almost all numbers have decimals in which each of the digits $0, \ldots, 9$ appears a tenth of the time. Understanding what this means would be the aim of this project. We can give a hint of what is involved by sketching a result concerning numbers whose decimals do not contain a 4. Readers are invited to try to fill in the details, perhaps in a tutorial context. If we consider just the numbers in the interval from 0 and 1, those without a 4 in the first decimal place occupy only $9/10$ of the interval. Of these numbers, those also without a 4 in the second decimal place occupy only $9/10$ of the $9/10$. If we continue this reasoning we find that those numbers without a 4 in the first n places occupy a set of intervals of total length $(9/10)^n$. The limit of this is zero, and in this sense we can say that the measure of the set of numbers without a 4 is zero, and so almost all decimal numbers have a 4 somewhere.

Answers to Exercises

We have given here answers to most of the end of section exercises, but not to end of chapter exercises, as tutors may wish to use some of those for assessment purposes. Answers rather than full solutions are provided. It is hoped that appropriate proofs or worked examples will enable full, well written solutions to be reconstructed. The abbreviated style used is therefore not to be taken as a model of how to write mathematics. In a few cases, for example where readers are asked to write short computer programs, or write detailed proofs similar to those in the text, no answers are given.

EXERCISES 1.2

1. (i) $\{2, 3, 4, 5, 6, 7, 8\}$, (ii) $\{-4, -3, -2, -1, 0, 1, 2, 3, 4\}$,
    (iii) $\{11, 13, 17, 19, 23, 29, 31, 37, 41, 43, 47, 53, 59, 61, 67, 71, 73, 79, 83, 89, 97\}$,
    (iv) $\{1, -2, 3\}$, (v) $\{0, 1\}$.

2. (i) $\{t : t \in \mathbb{Z} \text{ and } 0 < |t| \leq 3\}$, (ii) $\{m : m = 3k - 2 \text{ where } k \in \mathbb{N}\}$,
    (iii) $\{-a^2 : a \in \mathbb{N}\}$, (iv) $\{y : y = 10^{-k}, k \in \mathbb{N}, \text{ written as a decimal}\}$,
    (v) $\{v : v \text{ is a vowel in the English alphabet}\}$.

3. (i) False, (ii) True, (iii) False, (iv) True, (v) False, (vi) False.

EXERCISES 1.3

1. In each case the contrapositive is written first, followed by the converse.

    (i) If m is a rational number then $m^2 \neq 10$. If m is not a rational number then $m^2 = 10$. Converse false.

    (ii) If $x^2 - 5x + 6 \neq 0$ then $x \neq 2$. If $x^2 - 5x + 6 = 0$ then $x = 2$. Converse false.

    (iii) If $\sin \theta \neq 0$ then $\theta \neq \pi$. If $\sin \theta = 0$ then $\theta = \pi$. Converse false.

    (iv) If $\int_0^1 f(x)\, dx \leq 0$ then $f(x) \not> 0$ for $0 \leq x \leq 1$. If $\int_0^1 f(x)\, dx > 0$ then $f(x) > 0$ for $0 \leq x \leq 1$. Converse false.

    (v) If $f(x)$ does not have a local maximum at $x = a$ then $f'(a) \neq 0$ or $f''(a) \geq 0$. If $f(x)$ has a local maximum at $x = a$ then $f'(a) = 0$ and $f''(a) < 0$. Converse false.

2. Sketch proofs given; readers to fill in details.

    (i) Contrapositive: m even implies m^3 even. Let $m = 2k$, then $m^3 = 8k^3$, i.e. even.

    (ii) Use the formula for the sum of an arithmetic progression in the contrapositive.

    (iii) Draw graphs of the quadratic functions.

(iv) Consider the possible remainders when x^2 is divided by 4 in the two cases of x even and x odd respectively. Then consider possible remainders for $x^2 + y^2$.

(v) The contrapositive is a statement of the theorem that the angle in a semicircle is a right-angle.

EXERCISES 1.4

1. (i) $\forall x \in \mathbb{Z}, x \in \mathbb{Q}$. True. (ii) $\exists t \in \mathbb{Q}, -1 < t < 1$. True.
   (iii) $\forall x, 2x^2 - 5x + 3 = 0 \Rightarrow x \in \mathbb{Z}$. False. (iv) $\forall y \in \mathbb{Q}, y < y^2$. False.
   (v) $\exists x \in \mathbb{R}, \sin x = -2$. False.

2. (i) Any real number whose cosine is equal to zero is less than 2π in magnitude. False. $\exists x \in \mathbb{R}, \cos x = 0$ and $|x| \geq 2\pi$. (ii) There is a rational number satisfying the equation $23q = 78$. True. (iii) If h is 1, 2 or 3 then $h^3 - h + 7$ is positive. True. (iv) Whenever $\cos \pi x$ is zero, x is a rational number. True. (v) Every integer is either negative or has a real square root. True.

EXERCISES 1.5

1. (i) $\exists m \in \mathbb{Z}, \exists n \in \mathbb{Z}, 2m + 1 = 3^n$. True. (ii) $\forall s \in \mathbb{Q}, s > 0 \Rightarrow \exists t \in \mathbb{Q}, 0 < t < s$. True. (iii) $\forall a \in \mathbb{R}, \exists x \in \mathbb{R}, \cos(ax) = 0$. False. (iv) $\forall x \in \mathbb{R}, \exists n \in \mathbb{Z}, x < n < 2x$. False. (v) $\forall k \in \mathbb{R}, \exists x \in \mathbb{R}, kx = 1$. False. (vi) $\forall a \in \mathbb{R}, \exists x, \exists y, x \neq y$ and $x^2 = y^2 = a$. False.

2. (i) Given any real number x, there is a solution y of the equation $x + y = 0$. True. (ii) There is a real number y which, whatever number x is added to it, gives zero. False. Negation: $\forall y \in \mathbb{R}, \exists x \in \mathbb{R}, x + y \neq 0$. (iii) For every real number t and for every positive integer n, nt is greater than t. False. Negation: $\exists t \in \mathbb{R}, \exists n \in \mathbb{N}, nt \leq t$. (iv) There are integers a and b, with a positive, satisfying the equation $a^2 - b^2 = 3$. True. (v) Corresponding to any real number, we can find another real number so that the sum of the two is less than all possible real numbers. False. Negation: $\exists u \in \mathbb{R}, \forall v \in \mathbb{R}, \exists w \in \mathbb{R}, u + v > w$.

EXERCISES 1.6

1. Answers are given to indicate which of the three properties (R,S,T) are satisfied. (i) None (remember sisters). (ii) R,S,T. (Ignoring any ambiguities for places having a county boundary going through the middle.) An example of an equivalence class would be the list of all towns and villages in Hampshire. (iii) R,S,T. An equivalence class is a set of all translates of one triangle. (iv) S only. (v) R,S,T. An equivalence class is a set of all functions of the form $f(x) + c$, where f is a fixed function and c ranges over all real numbers. (vi) R,S,T. The equivalence classes are circles, centred at the origin.

EXERCISES 2.2

In each case we have sketched the anchor step, and given just the algebra involved in the inductive step. Readers should formulate a statement of the inductive hypothesis.

1.   (i) Anchor: LHS $=$ RHS $= 1$ when $n = 1$.
   Inductive step: $k^2 + 2(k+1) - 1 = (k+1)^2$.

   (ii) Anchor: LHS $=$ RHS $= 1$ when $n = 1$
   Inductive step: $k(k+1)/2 + (k+1) = (k+1)(k+2)/2$.

   (iii) Anchor: LHS $=$ RHS $= 1$ when $n = 0$.
   Inductive step: $(1 - x^{k+1})/(1 - x) + x^{k+1} = (1 - x^{k+2})/(1 - x)$.

2. Anchor: LHS $=$ RHS $= 1 + x$ when $n = 1$.
   Inductive step: $(1 + x)^{k+1} = (1 + x)(1 + x)^k \geq (1 + x)(1 + kx) = 1 + (k+1)x + kx^2 \geq 1 + (k+1)x$.

3. Anchor: When $n = 1$, $5^n + 2.3^{n+1} + 1 = 24$.
   Inductive step: $5^{k+1} + 2.3^{k+2} + 1 = 5.5^k + 6.3^{k+1} + 1 = (4.5^k + 4.3^{k+1}) + (5^k + 2.3^{k+1} + 1)$. Both brackets are divisible by 4.

4. Anchor: $H = 12$ is the smallest value.
   Inductive step: $(k+1)! = (k+1)k! \geq (k+1)5^k \geq 5^{k+1}$ since $k \geq 12$.

5. The hypotenuse of the nth triangle has length $(1/\sqrt{2})^{n-2}$. The area of the nth triangle is $(1/2)^n$. The total area of the first n triangles is $1 - (1/2)^n$.

6. $S(2n) = -n$; $S(2n + 1) = n + 1$. $S(n) = (-1)^{n+1}\frac{1}{2}(n + \frac{1}{2}(1 - (-1)^n))$.

EXERCISES 2.5

1.   (i) h.c.f. $= 3 = 87(5 - 24m) + 72(29m - 6)$.

   (ii) h.c.f. $= 29 = 1073(5m - 2) + 145(15 - 37m)$.

   (iii) h.c.f. $= 1 = 7537(8039m - 3443) + 8039(3228 - 7537m)$.

2. Let $c = kd$. Then $d = ax_1 + by_1$ for some x_1, y_1 so $c = a(kx_1) + b(ky_1)$. If $c = dq + r$ $(0 \leq r < d)$ then $ax + by = (ax_1 + by_1)q + r$. So $r = a(x - x_1q) + b(y - y_1q)$. Now $r < d$ and so we must have $r = 0$, so $d|c$.

3.   (i) The highest common factor of 301 and 84 is 7, which is not a divisor of 5. No solutions.

   (ii) h.c.f.$(345, 735) = 15|60 : 345(49m - 68) + 735(32 - 23m) = 60$.

   (iii) h.c.f.$(87, 53) = 1|13 : 87(53m - 182) + 53(299 - 87m) = 13$.

4. $d = ax + by$ for some x, y so $nd = nax + nby$. Now $nax + nby = n(ax + by) < nd$ only if $ax + by < d$ since $n > 0$. But d is the smallest positive integer combination of a, b so nd is the smallest positive integer combination of na, nb.

EXERCISES 2.6

1. 100/10/11110010110.

3. Take the digits in pairs. The following example illustrates the procedure. Consider 2311322 in base four. We group this as 2|31|13|22. Consider each pair.

$$22 \sim 2 \times 4 + 2 = 10_{TEN} = A_{HEX}.$$

$$13 \sim 1 \times 4^3 + 3 \times 4^2 = (1 \times 4 + 3).4^2 = 7 \times 16.$$

$$31 \sim (3 \times 4 + 1).4^4 = 13_{TEN} \times 16^2 = D_{HEX}.16^2.$$

Finally, $2 \sim 2.4^6 = 2.16^3$. So 2311322 in base four converts into the hexadecimal representation 2D7A. Once the pairing mechanism is understood we do not need to write down the intermediate steps as we have done here. In fact there are only 16 possible pairs, so we could simply compile a conversion table for these directly into hexadecimal and then refer to that.

4. Following the procedure in Example 11, we find that we need the six times table in base eight in order to perform the divisions. Once we have complied this we do not need to refer to base ten.

6. $144 = m^2 + 4m + 4$ in base m. This is $(m + 2)^2$ which is 12^2 in base m.

EXERCISES 3.1

1. $px = q \Rightarrow (px)y = qy \Rightarrow p(xy) = qy \Rightarrow p(xy) = p \Rightarrow p^{-1}(p(xy)) = p^{-1}p \Rightarrow (p^{-1}p)(xy) = p^{-1}p \Rightarrow 1(xy) = 1 \Rightarrow xy = 1.$

2. (a) $qx = p, sy = r \Rightarrow qsx = ps, qsy = qr \Rightarrow qs(x - y) = ps - qr.$

   (b) $qx = p, sy = r \Rightarrow qsxy = pr.$

   (c) $qx = p, sy = r \Rightarrow s^{-1}y^{-1} = r^{-1} \Rightarrow qs^{-1}xy^{-1} = pr^{-1} \Rightarrow qr(xy^{-1}) = ps.$

3. Suppose $\forall a, a + z_1 = a + z_2 = a$. $\exists x, a + x = z_1 \Rightarrow a + x + z_1 = a + x + z_2 \Rightarrow z_1 + z_1 = z_2 \Rightarrow z_1 = z_2.$

   Suppose $\forall a \neq z, a \times e_1 = a \times e_2 = a$. $\exists y, a \times y = e_1 \Rightarrow a \times y \times e_1 = a \times y \times e_2 \Rightarrow e_1 \times e_1 = e_1 \times e_2 \Rightarrow e_1 = e_2.$

4. $(a \times b) \times (b^{-1} \times a^{-1}) = a \times (b \times b^{-1}) \times a^{-1} = a \times e \times a^{-1} = a \times a^{-1} = e$ using M2, M3 and M4.

EXERCISES 3.2

1. $a_1 b_2 = a_2 b_1, c_1 d_2 = c_2 d_1 \Rightarrow a_1 b_2 d_1 d_2 = a_2 b_1 d_1 d_2, c_1 d_2 b_1 b_2 = c_2 d_1 b_1 b_2$. Subtract and factorize. $(a_1 d_1 - b_1 c_1) b_2 d_2 = (a_2 d_2 - c_2 d_2) b_1 d_1.$

   $a_1 b_2 = a_2 b_1, c_1 d_2 = c_2 d_1 \Rightarrow a_1 b_2 c_1 d_2 = a_2 b_1 c_2 d_1 \Rightarrow (a_1 c_1)(b_2 d_2) = (a_2 c_2)(b_1 d_1).$

   $a_1 b_2 = a_2 b_1, c_1 d_2 = c_2 d_1 \Rightarrow a_1 b_2 c_2 d_1 = a_2 b_1 c_1 d_2 \Rightarrow (a_1 d_1)(b_2 c_2) = (a_2 d_2)(b_1 c_1).$

EXERCISES 3.3

1. $89/55 = [1; 1, 1, 1, 1, 1, 1, 1, 2]$　　$1101011/1001010 = [1; 10, 100, 1000]$
   $1393/972 = [1 : 2, 3, 4, 5, 6]$　　$6961/972 = [7; 6, 5, 4, 3, 2]$
   $169/70 = [2; 2, 2, 2, 2, 2].$

3. In the first case the first seven convergents are $1/1; 2/1; 5/3; 7/4; 12/7; 55/32;$
    $67/39;$ giving e $\approx 2.7179\ldots$. The first two decimal places are correct. In the
    second case the first seven convergents are $0/1; 1/1; 6/7; 61/71; 860/1001;$
    $15541/18089; 342762/398959;$ giving e $\approx 2.718286\ldots$. The first five decimal
    places are correct.

EXERCISES 4.1

1. $0 < a < b \Rightarrow 0 < a.a < a.b; 0 < a < b \Rightarrow 0 < a.b < b.b.$ Transitivity gives
    $a.a < b.b.$

2. $-1 < 3$ and $(-1)^2 < 3^2.$ $-4 < -2$ but $(-4)^2 > (-2)^2.$

3. $a < b \Rightarrow a + c < b + c; c < d \Rightarrow c + b < d + b.$ Transitivity gives $a + c < b + d.$

4. $3 < 7, 2 < 4$ and $3 - 2 < 7 - 4; 3 < 4, 2 < 7$ but $3 - 2 > 4 - 7.$

5. $a > 0, (1/a) < 0 \Rightarrow a.(1/a) < 0,$ i.e. $1 < 0,$ contradiction. So if $a > 0$ the
    reciprocal of a is also positive. Now $a < b \Rightarrow a.(1/a) < b.(1/a) \Rightarrow$
    $1 < b.(1/a) \Rightarrow (1/b) < (1/b).b.(1/a) \Rightarrow (1/b) < 1.(1/a) = (1/a).$

EXERCISES 4.2

1. (i) $x < 2 - \sqrt{5}$ or $x > 2 + \sqrt{5}.$ (ii) $x = 2.$ (iii) $2 - \sqrt{6} < x < 2 + \sqrt{6}.$ (iv) No
    solutions.

2. $x < 0$ or $x > 1.$

3. $x < 3.7944$ rounded correct to four decimal places.

EXERCISES 4.3

1. For Exercise 1, we complete the square, so $x^2 - 4x + 2 = (x - 2)^2 - 2.$ The four
    inequalities then become $(x - 2)^2 > 5; (x - 2)^2 \leq 0; (x - 2)^2 < 6; (x - 2)^2 < -1;$
    giving the solutions quoted for Exercises 4.2.

    For the inequality of Exercise 2, we must consider two cases, $2x - 1 > 0$ and
    $2x - 1 < 0.$ Multiplying by $2x - 1$ when positive preserves the inequality, and
    when rearranged gives $x > 0$ and $x > 1,$ so giving $x > 1.$ Multiplying by $2x - 1$
    when negative reverses the inequality, and when rearranged gives $x < 0$ and
    $x < 1.$ Since in this case $x < \frac{1}{2}$ this gives $x < 0.$ So the total solution is $x < 0$ or
    $x > 1.$

2. Since $x^2 + 2$ is always positive we can cross multiply. The solution is $-1 < x < 0$
    or $x > 5.$

EXERCISES 4.4

1. (i) $x < -2$ or $1 < x < 3.$

    (ii) $-3 < x < 2$ or $2 < x < 3.$ Note that when $x = 2$ the left-hand side is equal
    to zero.

    (iii) $x < -2$ or $x > 3.$

(iv) ... $-\dfrac{7\pi}{2} < x < -\dfrac{5\pi}{2}, -\dfrac{3\pi}{2} < x < -3, -\dfrac{\pi}{2} < x < \dfrac{\pi}{2},$

$\dfrac{3\pi}{2} < x < \dfrac{5\pi}{2}, \dfrac{7\pi}{2} < x < \dfrac{9\pi}{2}, \dots$

(v) $0 < x < 1, \pi < x < 2\pi, 3\pi < x < 4\pi, 5\pi < x < 6\pi, \dots.$

EXERCISES 4.5

1. Taking logarithms gives $x < x^2 \ln 2$, giving $x < 0$ or $x > 1/\ln 2$.

2. The cube function is increasing, so the solution is the same as the inequality without the cube. This gives $x > 13$.

3. One might say that since the logarithmic function is increasing, and $\ln 1 = 0$, the inequality becomes that of Example 7, so that $x < -3$ or $x > -2$. This ignores the fact that one cannot have the logarithm of a negative number. So we must have $(2 - x)/(12 + 4x) > 0$, giving $-3 < x < 2$. So the solution of the inequality is $-2 < x < 2$.

EXERCISES 5.1

2. $r^2 = 12 \Rightarrow (r/2)^2 = 3$. Now use the result of Exercise 1.

3. Suppose $(p/q)^3 = 4$, where p and q have no common factors. Then $p^3 = 4q^3$. $2|p^3 \Rightarrow 2|p$. (Prove this.) Let $p = 2a$, so $8a^3 = 4q^3$, giving $q^3 = 2a^3$. $2|q^3 \Rightarrow 2|q$. Contradiction.

EXERCISES 5.3

1. (i) g.l.b. $= 2$; min $= 2$; no l.u.b.; no max.

    (ii) g.l.b. $= 0$; min $= 0$; l.u.b. $= 1$; no max.

    (iii) g.l.b. $=$ min $= -1$; l.u.b. $=$ max $= 1$.

    (iv) g.l.b. $= 0$; no min; l.u.b. $= 3^{-1} + 5^{-1} =$ max.

    (v) Unbounded above and below.

    (vi) g.l.b. $= 1$; no min; l.u.b. $=$ max $= 2$.

    (vii) g.l.b. $=$ min $= 1$; l.u.b. $=$ max $= 3$.

    (viii) g.l.b. $= 2$; no min; l.u.b. $= 5$; no max.

2. $x > 3 \Rightarrow x^2 > 8$; $x < -3 \Rightarrow x^2 > 8$. So S is bounded and $x^2 < 8 \Rightarrow -3 < x < 3$. To show that S has no largest member follow Example 4. To show that it has no smallest member adapt Example 4 to deal with negative numbers.

3. Let $l = $ l.u.b.(S). (i) $x \in S \Rightarrow x \leq l$. $x \in T \Rightarrow -x \in S \Rightarrow x \geq -l$. So T is bounded below. (ii) Let $\beta > 0$. $\exists x \in S, x > l - \beta$. So $-x < -l + \beta$, so we have found a member of T less than $-l + \beta$. Thus $-l = $ g.l.b.(T).

4. Let $A = \{x : 2 \leq x \leq 4\}$, $B = \{x : -3 \leq x \leq 1\}$. l.u.b.$(A) - $ l.u.b.$(B) = 4 - 1 = 3$. But $4 - (-3) = 7 \in D$, so $3 \neq$ l.u.b.(D).

5. Let $A = \{x : -3 \le x \le -1\}$, $B = \{x : -4 \le x \le -2\}$. Then
l.u.b.$(A) \times$ l.u.b.$(B) = -1 \times -2 = 2$. But $-3 \times -4 = 12$ belongs to the product
set.

6. If a is any member of S then $\forall y \in T, y \ge a$. So T is bounded below. Let
$l =$ l.u.b.(S). Then $l \in {}'T$. If $m < l$ then m is not an upper bound for S and so
$m \notin T$. Thus, l is the smallest member of T, which is therefore bounded below
and contains its g.l.b.

EXERCISES 5.4

1. 2 lies between the two numbers given.

2. $\dfrac{2}{3} + \dfrac{1}{12\sqrt{2}}$ lies between the two numbers given.

3. (i) True. (ii) Sometimes false, e.g. $\sqrt{2} + (1 - \sqrt{2}) = 1$. (iii) True. (iv) False only
if the rational is zero. (v) Sometimes false, e.g. $\sqrt{2} \times \sqrt{18} = 6$.

EXERCISES 6.2

1. (i) $2 + 34$i; (ii) $(7/65) + (4/65)$i; (iii) $-(8/25) - (19/25)$i.

2. $|z_1 z_2|^2 = (x_1 x_2 - y_1 y_2)^2 + (x_1 y_2 + x_2 y_1)^2$. Multiplying out, collecting terms and
factorizing gives $(x_1^2 + y_1^2)(x_2^2 + y_2^2) = |z_1|^2 |z_2|^2$. Now take square roots. This
gives an anchor, for $n = 2$. We use this case for the inductive step.
$|z_1 \ldots z_n z_{n+1}| = |z_1 \ldots z_n||z_{n+1}| = |z_1||z_2| \ldots |z_n||z_{n+1}|$.
$|1/z| = |(x - y\text{i})/(x^2 + y^2)| = (x^2 + y^2)/(x^2 + y^2)^2 = 1/(x^2 + y^2) = 1/|z|$.

3. $(z^*)^* = (x - y\text{i})^* = x - (-y)\text{i} = x + y\text{i} = z$.

4. $|z| = x^2 + (-y)^2 = x^2 + y^2 = |z|$.

5. $(1 + \text{i})^2 = 2\text{i}$ so $(1 + \text{i})/\sqrt{2}$ is a square root of i. The other one is $-(1 + \text{i})/\sqrt{2}$.
The square roots of $-\text{i}$ are $\pm(1 - \text{i})/\sqrt{2}$.

6. $z^2 - 2z - \text{i} = (z - 1)^2 + 1 - \text{i} = 0$, so $z = 1 \pm \sqrt{-1 + \text{i}}$.

7. $|z_1| = |(z_1 - z_2) + z_2| \le |z_1 - z_2| + |z_2|$. So $|z_1 - z_2| \ge |z_1| - |z_2|$, and similarly
$|z_1 - z_2| \ge |z_2| - |z_1|$.

EXERCISES 6.4

1. $|1 + \text{i}| = \sqrt{2}$; $\arg(1 + \text{i}) = \pi/4$; $|-\text{i}| = 1$; $\arg(-\text{i}) = -\pi/2$; $|1 - \text{i}\sqrt{3}| = 2$;
$\arg(1 - \text{i}\sqrt{3}) = -\pi/3$; $|-2| = 2$; $\arg(-2) = \pi$; $|\sqrt{3} - \text{i}| = 2$;
$\arg(\sqrt{3} - \text{i}) = -\pi/6$.

2. (i) Perpendicular bisector of the segment joining the points i and 1.

(ii) Circle, centre -1, radius 2.

(iii) The line $x = -3$ (parallel to the y-axis).

(iv) The half-plane below the line $y = 1$, and including the line itself.

3. $z_1 = -1 + i\sqrt{3} = 2(\cos(2\pi/3) + i\sin(2\pi/3))$;
   $z_2 = -1 + i = \sqrt{2}(\cos(3\pi/4) + i\sin(3\pi/4))$. $2\pi/3 + 3\pi/4 = 17\pi/12 \sim -7\pi/12$.
   So $\arg(z_1 z_2) = -7\pi/12$.

4. $1/z = (1/r)(\cos\theta - i\sin\theta)$.

5. $zz^* = r(\cos\theta + i\sin\theta)r(\cos\theta - i\sin\theta) = r^2(\cos^2\theta + \sin^2\theta) = r^2 = |z|^2$.

6. $z^{-n} = 1/z^n = 1/(r^n(\cos(n\theta) + i\sin(n\theta))) = (1/r^n)(\cos(-n\theta) + i\sin(-n\theta))$.

EXERCISES 6.5

1. $(e^{i\theta})^* = (\cos\theta + i\sin\theta)^* = \cos\theta - i\sin\theta = \cos(-\theta) + i\sin(-\theta) = e^{-i\theta}$.

2. Similar to Example 7 with appropriate changes of sign.

3. $|e^{i\theta}|^2 = \cos^2\theta + \sin^2\theta = 1$.

4. $e^{i\theta} = \cos\theta + i\sin\theta$; $e^{-i\theta} = \cos\theta - i\sin\theta$. Adding the two equations gives the first result, and subtracting gives the second.

EXERCISES 6.6

1. The fourth roots of unity are 1, $e^{i\pi/2}$; $e^{i\pi}$; $e^{i3\pi/2}$; i.e. 1; i; -1; $-i$.

2. $8i = 8e^{i\pi/2}$. The cube roots are $2e^{i(\pi/6 + 2n\pi/3)}, n = 0, 1, 2$, i.e. $2e^{i\pi/6}$; $2e^{i5\pi/6}$; $2e^{i9\pi/6}$.

3. Let $w_1 = e^{i2\pi/3}$; $w_2 = e^{i4\pi/3} = e^{-i2\pi/3} = w_1^*$; $w_1^2 = e^{i4\pi/3} = w_2$;
   $w_2^2 = e^{i8\pi/3} = e^{i2\pi/3} = w_1$; w_1 and w_2 are reflections of one another in the real axis. Together with the point 1 they form an equilateral triangle.

4. If $|z| = 1$ and $x = -\frac{1}{2}$ then $y^2 = 3/4$ so $y = \pm\sqrt{3}/2$. $-\frac{1}{2} \pm \frac{1}{2}\sqrt{3} = e^{\pm i2\pi/3}$. These are the two non-real cube roots of unity.

5. $z^6 - 1 = (z - 1)(z + 1)(z^2 - z + 1)(z^2 + z + 1)$.

6. The primitive 8th roots are $e^{i2\pi k/8}$ $(k = 1, 3, 5, 7)$.

EXERCISES 7.2

1. $x_{n+1} = (x_n + 5/x_n)/2$; $x_1 = 2$.

2. (i) $x = (x^3 - 4)/6$; $x = \sqrt[3]{6x + 4}$ etc.
   (ii) $x = \sqrt{\sin x}$; $x = (\sin x)/x$ etc.
   (iii) $x = \ln(3x^2 - 2)$; $x = \sqrt{(e^x + 2)/3}$ etc.

3. Suppose the angle subtended by the chord is $2x$. Then considerations of area lead to the equation $x = (\pi/3) + (\sin 2x)/2$. Solving numerically gives $x = 1.3026628\ldots$ and $OA = 0.2649321\ldots \times$ radius.

EXERCISES 7.3

1. $|b_n - |l|| = ||a_n| - |l|| \leq |a_n - l|$ by Exercise 7 of §6.2. So $|a_n - l| < \epsilon \Rightarrow |b_n - |l|| < \epsilon$. If $a_n = (-1)^n$ then $b_n = 1$ for all n so $\lim(b_n) = 1$, but (a_n) has no limit.

3. (i) 1; (ii) ∞; (iii) oscillates unboundedly; (iv) 0; (v) 0 if $|a| < |b|$, oscillates unboundedly if $a/b < -1$, ∞ if $a/b > 1$, 1 if $a = b$, oscillates boundedly if $a = -b$; (vi) 0; (vii) 0; (viii) 1; (ix) 0; (x) ∞; (xi) ∞; (xii) 1; (xiii) 1/3; (xiv) 1; (xv) a; (xvi) ∞; (xvii) oscillates boundedly; (xviii) ∞; (xix) 1 if $\sin\theta = 1$, oscillates boundedly if $\sin\theta = -1$, otherwise zero; (xx) no limit unless θ is a multiple of 2π.

4. (i) $a_n = n$; $b_n = n^2$, (ii) $a_n = n$; $b_n = n$, (iii) $a_n = n^2$; $b_n = n$,
(iv) $a_n = n + 2$; $b_n = n$, (v) $a_n = n^2 + (-1)^n n$; $b_n = n^2$, (vi) $a_n = n^2$; $b_n = n$,
(vii) $a_n = n$; $b_n = n^2$, (viii) $a_n = 3n$; $b_n = n$.

EXERCISES 7.5

The proofs for these exercises are similar to those of Examples 8 and 9.

1. $e_{n+1}/e_n = (1 + a_n + a_n^2)/3 \to 1$ so convergence will be slow.

   If $a_1 < -2$ the sequence decreases and diverges. If $a_1 > 1$ the sequence increases and diverges. If $a_1 = -2$ then $a_n = -2$ for all n, so $\lim(a_n) = -2$.

2. (i) If $a_1 < 0$ then eventually some term of the sequence will be positive. From that point on the terms alternate either side of $1 + \sqrt{2}$. The behaviour is similar to the sequence in Example 9. The limit is $1 + \sqrt{2}$.
   $\lim(e_{n+1}/e_n) = (\sqrt{2} - 1)/(\sqrt{2} + 1) \approx 0.17$.

   (ii) For any $a_1 \neq 0$ the sequence converges with limit 1. Eventually some term of the sequence will exceed 1 and from then onwards the sequence is decreasing. $\lim(e_{n+1}/e_n) = 1$.

3. $a_{n+1} = \sqrt{2 + a_n}$; $a_1 = \sqrt{2}$. The limit of the sequence is 2. The inequality tells us that accuracy is at least doubled at each stage.

EXERCISES 7.6

1. $||a_n| - 0| = ||a_n|| = |a_n| = |a_n - 0|$. So $|a_n - 0| < \epsilon$ if and only if $||a_n| - 0| < \epsilon$. $|a_n| = r_n$ so $\lim(a_n) = 0$ if and only if $\lim(r_n) = 0$.

2. (i) $|a_n| = |0.7 + 0.7i|^n = (0.98)^{n/2} \to 0$. (ii) $|a_n| = |0.8 + 0.8i|^n = (1.28)^{n/2} \to \infty$.
   (iii) $|a_n| = |0.8 + 0.6i|^n = 1 \nrightarrow 0$. $\arg(0.8 + 0.6i) = \tan^{-1}(4/3)$, so the sequence has no limit. (iv) $|a_n| = 1$ for all n. $\arg(a_n) = \alpha/n \to 0$, so $\lim(a_n) = 1$.

EXERCISES 8.1

1. $$s_n = 1 + 2r + 3r^2 + \ldots + nr^{n-1},$$
$$rs_n = \qquad r + 2r^2 + \ldots + (n-1)r^{n-1} + nr^n,$$
$$s_n(1 - r) = 1 + r + r^2 + \ldots + r^{n-1} - nr^n = (1 - r^n)/(1 - r) - nr^n, \quad (r \neq 1),$$
$$s_n = (1 - r^n)/(1 - r)^2 - nr^n/(1 - r).$$

For $|r| < 1$, $\lim(s_n) = 1/(1 - r)^2$. For $|r| \geq 1$ we have divergence.

2. $\dfrac{j}{(j+4)(j+5)(j+6)} = -\dfrac{2}{j+4} + \dfrac{5}{j+5} - \dfrac{3}{j+6}$. The sum cancels down to

$-\dfrac{2}{5} + \dfrac{3}{6} + \dfrac{2}{n+5} - \dfrac{3}{n+6}$. The limit is $-\dfrac{2}{5} + \dfrac{3}{6} = \dfrac{1}{10}$.

3.

$$\sum_{k=1}^{n} \cos(\alpha + k\theta) = \cos\left(\alpha + \frac{n+1}{2}\theta\right) \frac{\sin(n\theta/2)}{\sin(\theta/2)}$$

$$\sum_{k=1}^{n} \sin(\alpha + k\theta) = \sin\left(\alpha + \frac{n+1}{2}\theta\right) \frac{\sin(n\theta/2)}{\sin(\theta/2)}$$

$$\sum_{k=1}^{n} (-1)^{k+1} \cos(\alpha + k\theta) = -\cos\left(\alpha + \frac{(n+1)(\theta+\pi)}{2}\right) \frac{\sin(n(\theta+\pi)/2)}{\cos(\theta/2)}$$

$$\sum_{k=1}^{n} (-1)^{k+1} \sin(\alpha + k\theta) = -\sin\left(\alpha + \frac{(n+1)(\theta+\pi)}{2}\right) \frac{\sin(n(\theta+\pi)/2)}{\cos(\theta/2)}.$$

For the second pair of series use $(-1)^{k+1} = e^{i(k+1)\pi}$ to obtain a geometric series. In all four the summation and identity from Example 2 can be used.

EXERCISES 8.2

1. (i) D (comparison), (ii) C (comparison), (iii) D (comparison), (iv) D (evaluate partial sum), (v) C (comparison), (vi) C (comparison), (vii) C (geometric series), (viii) D (*n*th term does not tend to zero), (ix) D (*n*th term does not tend to zero), (x) C (Leibniz's test).

2. (i) C, (ii) D, (iii) C, (iv) C, (v) D, (vi) D.

3. $\displaystyle\sum_{r=1}^{\infty} \frac{(r+1)^2}{r!} = \sum_{r=1}^{\infty} \frac{r(r-1) + 3r + 1}{r!}$

$$= \sum_{r=1}^{\infty} \frac{r(r-1)}{r!} + \sum_{r=1}^{\infty} \frac{3r}{r!} + \sum_{r=1}^{\infty} \frac{1}{r!}$$

$$= 0 + 1 + \sum_{m=1}^{\infty} \frac{1}{m!} + 3 + 3\sum_{m=1}^{\infty} \frac{1}{m!} + \sum_{m=1}^{\infty} \frac{1}{m!}$$

$$= 0 + 1 + (e-1) + 3 + 3(e-1) + (e-1) = 5e - 1.$$

EXERCISES 8.3

1. $\displaystyle\int_3^k \frac{dr}{r \ln r (\ln(\ln r))^p} = \left[\frac{(\ln(\ln r))^{1-p}}{1-p}\right]_3^k \quad (p \neq 1)$

$\to 0$ as $k \to \infty$ if $p > 1$;

$\to \infty$ as $k \to \infty$ if $p < 1$.

$\displaystyle\int_3^k \frac{dr}{r \ln r \ln(\ln r)} = [\ln(\ln(\ln r))]_3^k \to \infty$ as $k \to \infty$.

So the series converges if $p > 1$ and diverges if $p \leq 1$.

2. (i) $\displaystyle\sum_{n=m+1}^{\infty} n^{-5/3} \le \int_m^{\infty} x^{-5/3}\, dx = \frac{3}{2}\, m^{-2/3} < 10^{-4}$ for

$$m > \left(\frac{3}{2} \times 10^4\right)^{3/2} \approx 1,837,117.$$

(ii) $\displaystyle\sum_{n=m+1}^{\infty} \frac{1}{n(\ln n)^4} < \int_m^{\infty} \frac{dx}{x(\ln x)^4} = \frac{1}{3(\ln m)^3} < 10^{-4}$ if

$$m > \exp(\sqrt[3]{10^4/3}) \approx 3,072,542.$$

EXERCISES 8.5

1. In each case applying the ratio test gives a ratio limit of zero for all values of z.

2. (i) $R = 1$; D, D (*n*th term does not tend to zero when $z = \pm 1$),

   (ii) convergent for all z (it is the series for e^{2z}),

   (iii) $R = \frac{1}{2}$; C, C (standard series when $z = \pm\frac{1}{2}$),

   (iv) $R = 2$; D, D (*n*th term does not tend to zero when $z = \pm 2$),

   (v) $R = 1$; D, D (*n*th term does not tend to zero when $z = \pm 1$),

   (vi) $R = 1$; C when $z = 1$, D when $z = -1$,

   (vii) $R = 1$; D, D (*n*th term does not tend to zero when $z = \pm 1$),

   (viii) $R = 1$; C, C (standard series when $z = \pm 1$),

   (ix) $R = 1$; D, D (*n*th term does not tend to zero when $z = \pm 1$).

3. The ratio test gives $r = 1$. Use the results of Exercise 1 of §8.1 for the sum.

Index